MANKIND
BEYOND EARTH

MANKIND BEYOND EARTH

THE HISTORY, SCIENCE, AND FUTURE OF HUMAN SPACE EXPLORATION

CLAUDE A. PIANTADOSI

COLUMBIA UNIVERSITY PRESS   NEW YORK

COLUMBIA UNIVERSITY PRESS

PUBLISHERS SINCE 1893

NEW YORK   CHICHESTER, WEST SUSSEX

Library of Congress Cataloging-in-Publication Data
Piantadosi, Claude A.
   Mankind beyond Earth : the history, science, and future of human space exploration / Claude A. Piantadosi.
      p.   cm.
   Includes bibliographical references and index.
   ISBN 978-0-231-16242-5 (cloth : alk. paper)—ISBN 978-0-231-16243-2 (pbk. : alk. paper)—
ISBN 978-0-231-53103-0 (e-book)
   1. Astronautics—United States—Forecasting—Popular works.   2. Manned space flight—
History—Popular works.   3. Outer space—Exploration—Popular works.   I. Title.
   TL789.8.U5P53 2012
   629.4500973—dc23

                                                                                    2012017631

c 10 9 8 7 6 5 4 3
p 10 9 8 7 6 5 4 3 2

COVER ART: © CHESLEY BONESTELL, *EXPLORING MARS*, REPRODUCED COURTESY OF BONESTELL LLC.
COVER DESIGN: CHANG JAE LEE

Civilization is obliged to become spacefaring—not because of exploratory or romantic zeal, but for the most practical reason imaginable—staying alive.
CARL SAGAN

CONTENTS

Preface   xi

A SHORT INTRODUCTION TO THE SCIENCE OF
SPACE EXPLORATION   1

PART 1: HINDSIGHT AND FORESIGHT

1. MEN AND MACHINES   13
A House Divided   14
Robot Days   18
So Few Plausible Options   22

2. A SPACE LEXICON   25
The Thin Blue Line   28
The Science of Limits   30
Genes and Adaptation   33

Some Astronomical Concepts   36
The Tortoise and the Hare   40

3. THE EXPLORERS   45
People of Adventure   46
Polar Science, Space Science   49
The Hard-Shell Engineers   59
The Space Doctors   61

4. TWENTIETH-CENTURY SPACE   68
The Early Days   68
Project Apollo   71
The Shuttle Disasters   74
LEO and the Space Station   79

5. BACK TO THE MOON   88
Serene Selene   92
Water, Water Everywhere   95
It's Not Made of Green Cheese   99
Extravehicular Activity   105
The View from Earth   108

PART 2: A HOME AWAY FROM HOME

6. LIVING OFF THE LAND   117
Energy and Efficiency   119
Ecological Footprints   125
On Never Running Out of Air   127
Water and Food   133

**7. ROUND AND ROUND IT GOES . . .
WHERE IT STOPS, NOBODY KNOWS   140**
Recycling: Open or Closed, Hot or Cold?   141
The CO_2 Cycle   143
Making Room for the Jolly Green Giant   146
Micronutrients   149

8. BY FORCE OF GRAVITY   152
Allometry   153
Balance and Perception   155
Bone and Muscle   157
Weightlessness and the White Cell   165

9. THE COSMIC RAY DILEMMA   169
The ABCs of Cosmic Rays   172
The ALARA Principle   175
Radiation's Effects   181
Countermeasures   186

10. TINY BUBBLES   188
Destination Mars   188
Microbes in Space   192
Gas Leaks   195
The Age of the Astronaut   197
On Biological Clocks   199

PART 3: WHERE ARE WE GOING?

11. THE CASE FOR MARS   203
The Weather Forecast   207
Surface Time   211
Suits and Structures   214
One-Way Trips   216

12. BIG PLANETS, DWARF PLANETS, AND SMALL BODIES   219
The Truth About Asteroids   221
Oasis Ceres   223
Titan and the Galileans   225
Moons of the Ice Giants   231

13. NEW STARS, NEW PLANETS   236
On Leaving the Solar System   238
The Interstellar Medium   243

Changing Ourselves 245
Science Fact, Science Fiction 248

Notes 251
Bibliography and Additional Reading 253
Index 271

PREFACE

NASA's space shuttles retired in 2011 after an impressive thirty-year career. Two terrible accidents aside, the Space Shuttle Program's 135 missions epitomize the best of human spaceflight since the Apollo lunar landings of 1969 through 1972. The shuttle program produced many unforgettable highlights, but most important, without it we could have never built the International Space Station (ISS), a technological tour de force of which we all should be proud.

Not all the news, however, is good. The reasons to explore space first dawned on me as an eighth grader, when I carefully prepared my first science project on the Red Planet, and the urgency of those reasons has only increased. But in becoming a scientist, I was taught to see problems, and Houston, *we have a problem.* The shuttle program confirmed the high risk and cost of sending people into space, despite the attendant benefits, and the reality took the bloom off the rose long ago.

The shuttle was expensive, undependable, and overdue for the museum when the program ended, but we have no consensus on what's next for America's manned spaceflight program. Should we focus on our investment in the

ISS, intercept an asteroid, or go back to the Moon? And what about Mars? These are all figuratively and literally moving targets, and the decisions we must make are difficult ones.

The ISS was expensive to build, and it has been underutilized and undercapitalized. NASA understood this and managed to extend its working lifespan until 2020. Yet today, NASA has no heavy-lift launch capability. Thus we are now paying the Russian Federation to transport our people to the ISS. The 2011 deal pays the Russians 62.7 million dollars per seat to ferry, in 2014 and 2015, on Soyuz spacecraft, a dozen American astronauts to the ISS, and the test date for a new American launch vehicle has slipped to 2017 at the earliest. Confronted with these facts and figures, I hope you ask, "How did we arrive at this state of affairs?" rather than, "We farm out everything else, so why not space exploration?"

This situation is unprecedented. America has been the unquestioned leader in space since Project Apollo, but today we seem indifferent, at best, about our space program. In a world of virtual worlds, we can explore everything under the sun without ever leaving our Web browser, but cyberspace offers only what someone, someplace already knows—nothing more. The discovery of the unknown is the purpose of science, and therein lies real excitement. There is much to learn in space, and people (astronauts) can still discover things that machines (robots) cannot.

I don't want to give the wrong impression: robots are indispensable and will always go where we cannot. We need complementary goals for manned and unmanned space exploration, and both implemented at a price we can afford. What new space technologies are essential? Where should the government focus its investment in spaceflight? Which private-sector enterprises should be capitalized by federal dollars?

These are big questions at a time when a postrecession federal budget has spiraled out of control, government expenditures are huge and expanding, and our elected officials are hamstrung by conflicting priorities and political insecurities. Moreover, many politicians are skeptical of the ability of Americans to make informed judgments about science, and some influential people have become confused, indifferent, or even obstructionist. President Obama's speech at Cape Canaveral in 2010 opened some eyes when he cancelled Project Constellation and said, "I understand that some believe that we should attempt a return to the surface of the Moon first, as previously planned. But the simple fact is we have been there before."

Project Constellation was indeed in political and financial trouble, but the Moon is 1.5 times larger than North America, and we have only explored an area of it the size of Manhattan. Imagine if Isabella had decided not to send Columbus back to the New World because he had been there before! Mr. Obama has appeared to support human space exploration, but he has taken highly questionable advice in pointing us at near-Earth asteroids, under the guise of avoiding celestial Armageddon. Statistically, life on Earth being annihilated by an asteroid is far less likely than some of the terrible things we humans can already—and may very well—do to ourselves. Astronauts may not fly under the Stars and Stripes again until 2020, leaving just five years to prepare for an asteroid mission, and this policy has raised eyebrows for those people committed to the continuity of space exploration.

Most Americans are either unsure of what to make of this or are too preoccupied to notice. Indeed, public perception of our manned spaceflight program is fickle, and a certain amount of disinterest can be chalked up to our waning scientific literacy. According to the National Center for Education Statistics, in 2006 the average American high school student ranked seventeenth internationally in science knowledge, between Iceland and the Slovak Republic—and well below our Canadian neighbors. Perhaps it's no surprise that so few of us seem concerned about how this portion of our tax dollars are being spent.

In opinion polls, most Americans do view the space program favorably, at least until cost and risk are factored in, but most also do not see a compelling reason for a continued human presence in space—nor the value in the tax dollars spent on it. But even during the heady days of Apollo, only half of Americans felt that the money spent on lunar exploration was worthwhile. This is a curiously weak endorsement of an achievement that occupies a niche in history alongside the discovery of the New World. Some respectable scientists, too, question NASA's scientific returns on investment, but these critics rarely suggest realistic milestones as alternatives. Given all of this, there is good reason for a nonaligned but informed scientist to explain, in plain language, the fiscal, technological, and biological problems involved in pushing human exploration of space to the next level. In fairness, I am not fully disinterested; my center has worked on NASA contracts on and off for twenty years—but I am not a NASA man.

In writing this book, I chose a historical approach because it helps us see where we are, how we got here, where we might be headed, and, in particular, why things take so long. The thoughtful reader will see a way forward and, I hope, become enthused about scientific discovery and convinced that space exploration is worth the investment. It is a model for scientific progress and a part of human evolution.

This book emerged from a course entitled "Extreme Environments," which we have offered for more than ten years at the Duke University School of Medicine. However, this is not a medical textbook; it is written specifically with the general science reader in mind. It requires nothing more than a little introductory physics, chemistry, and biology, and a bit of persistence.

The course has covered many things, ranging from the origins of life on Earth to the limits of the depths of the oceans, the rarified atmosphere of the highest mountain peaks, the near vacuum of space, and the possibility of life on other planets. In examining these striking aspects of our world, my colleagues and I noticed that depth and breadth in these areas are hard to find. We accumulated a lot of information, and I have assembled and presented the best of it here in nontechnical language, to make it accessible to as many people as possible.

I also chose Carl Sagan's celebrated words as the epigraph for the book because it affixes a survival imperative to human space exploration. Whether or not you agree with Sagan, after only fifty years, human space exploration seems stuck in low Earth orbit, and his far-reaching vision seems to have been drawn into space's vacuum. To quote my friend and colleague "Q," an MIT-trained engineer and former NASA scientist: "Ultimately, we need a backup plan."

Regrettably, our bodies are not designed to live in space, and without the protective umbrella of our terrestrial atmosphere and magnetic field, it is an insuperable barrier. Space is a *something* in physics, but to us it is *nothing*—we die almost instantly from unprotected exposure. There is no atmosphere, no water, too little or too much gravity, and it is too hot or too cold. To make matters worse, it is shot through with lethal cosmic radiation, and safe havens are few and far between. In order to explore our Solar System, these issues must be confronted decisively, and I will present some ideas and solutions proposed by the experts, without getting mired in too many technological details.

This quest seeks to solve space's impenetrability by setting the horizon just right, melding the practicalities of space exploration with the constraints of human biology. In presenting the challenges, I have emphasized continuity and why good science and patience are the keys to unlocking the door to space. In short, I hope you will better appreciate why space exploration is important, why it is difficult, and why, as a member of a scientific society, you are vital to seeing it into the future.

Before embarking on this fascinating journey, I thank my editors at Columbia University Press, my colleagues and our students at Duke, and our many visiting scientists for their special insights, penetrating questions, and patience with my reductionist and sometimes constricted view of the world. I have done my best to blend the excitement of that boy learning all about Mars with the healthy skepticism of a lifelong scientist. I hope to kindle in you a sense of excitement for some of the great adventures still awaiting us as a nation. The fits and starts are inevitable, but space exploration will always teach us to live within our means—first and foremost, here on Earth.

MANKIND
BEYOND EARTH

A SHORT INTRODUCTION TO THE

SCIENCE OF SPACE EXPLORATION

Shortly after the opening of the Space Age, in the third quarter of the twentieth century, a dozen men famously went to the Moon and back, at levels of acceptable risk and certainly with all the "right stuff." However, by the time the last space shuttle was retired in 2011, the sum of all human experience in space totaled about the lifetime of one centenarian. Today, a few astronauts circle the planet for a few months at a time on the International Space Station (ISS), learning to live there. The ISS has been inhabited continuously since November 2000, but the crew battles a slew of problems. The cost has been astronomical; each shuttle mission cost 450 million to 1.5 billion dollars (Shiga 2011). In retrospect, by almost any metric, this investment did not yield its potential dividends.

Fifty years after we first entered space, our prospects there are still uncertain. In the United States, the White House's 2004 Vision for Space Exploration had raised the bar by proposing to return to the Moon by 2020 and reach Mars by 2035. Enthusiasm for the Moon-Mars initiative was limited, and its poster child, Project Constellation, was canceled before it ever got off the ground. Sentiments ran high among the stakeholders, and the coals of the post-Apollo debate on robotic versus human missions reignited. The dispute boiled down to money, and a scuffle broke out in 2006, when NASA began slashing its science programs to meet the goals of the initiative. Although the prospect of returning to the Moon and going to Mars fascinates everyone, many critics, including many physical scientists, have asked why we need to send *people*. This question needs an answer, but so do our space pioneers and our citizens hurt by the shuttle retirement and the cancellation of Moon-Mars.

Moon-Mars collapsed America's space program into a huge thirty-year package at great cost. Opponents thought such a massive venture with no scientific agenda was a bad idea. The initiative was immediately pounced on by the American Physical Society, an esteemed enclave of physicists that, while it favors human exploration, lobbies against the big-mission mentality. The big mission on everyone's mind, Mars, will be more dangerous and expensive than the first Apollo lunar landing was in 1969. To get to Mars, astronauts must function in deep space for years and will face huge challenges in doing so.

The criticisms of "big missions"—their high cost, high risk, and low return, not to mention that there may be higher priorities at home—do not invalidate an argument for human space exploration. In space science, as in all of science, the mission is to discover the unknown and collect information that improves how we do things. If done correctly, costs come down over the long run. We also have a record of big unmanned projects, too, like the Hubble Space Telescope, which is rarely disparaged for producing a wealth of perplexing astrophysical data. The protestor often assumes that the information acquired by such research is predictable a priori, but predictability takes experience, which is a posteriori. To explore is to travel, literally or metaphorically, for the purposes of discovery. And discovery entails the interpretive skills and resolution of the human mind—sharply honed by working closely with a problem. In this respect, space exploration fits easily into the larger context of science.

For much of the twentieth century, the world looked to America for the best in science and technology, but today we are in a battle to maintain our scientific preeminence. In 2012, U.S. spending on science and technology research still exceeded that of any other country, but the big nations in the Far East are steadily gaining ground. If this trend continues, it risks putting us off the global pace of science because progress is proportional to the time and money invested in research. Such investments can pay considerable economic dividends, which will also be endangered. If science in America falters, it will be felt quickly not only in the space sciences but also in biomedicine, information technology, and the physical sciences. Moreover, it will be reflected in further weakening of our economy.

Each year, the budgets of the National Science Foundation, the National Institutes of Health, and NASA must be rescued from the chopping block. Too many of our elected officials are oblivious to the power of scientific investment and to the consequences of cutting it. This obliviousness is often hidden behind economic distractions, but it reflects our low science literacy compared with other technological nations. Americans have traditionally been fascinated by new science and technology, but the rise of scientific illiteracy leads to confusion about how to evaluate new science—as well as a tendency to devalue it. Misunderstandings arise because everything offered under the guise of science is not science, and knowing the difference takes science literacy and common sense.

This is not too surprising. Even seasoned scientists can have trouble evaluating the merits of something totally new. Science is built on objective truth, and the politically popularized notion that there is no objective truth is more than specious; it engenders false dichotomies and pointless arguments. We cannot know everything, but science is a method of discovery. If you are not a full-time scientist, it is worth a minute to examine how this is true. If you are, please forgive the extra page or so that follows.

To scholars, science is the discipline by which incontrovertible core knowledge is discovered, validated, and expanded in the face of new ideas. Armed with core knowledge, also called science, an observer can quantify, predict, or implement a change in the state of an object or the local environment with a quantifiable assurance that the outcome will fit the information base. For instance, knowledge of the genetic code and the cell allows the molecular biologist to effect novel changes in life processes. When new facts don't fit, something is wrong with the hypothesis, the methods, or the paradigm, and it is incumbent on the observer to find out which of those is wrong, and why.

Science is defined by ideas that might be disproven by experiment but that nonetheless stand the test of time. An idea that can be validated by experiment is a theory, and ideas that cannot be tested are sometimes confused with theories. A theory is testable; the nontestable is metaphysics. Evolution and intelligent design have both been called theories, but evolution passes test after test: it is science. Intelligent design can never be tested by human beings.

These constraints make it easy to identify the two kinds of science. The first acquires core knowledge simply for the sake of understanding; the second acquires it as a means to an end. Scientists tend to split knowledge acquisition into basic and applied research, for example, physics and engineering. Good basic science, however, almost inevitably leads to applications, and applications often raise more basic questions, particularly if those applications fall short of their stated goals.

The use of science as the means to an end—good or evil—has long given pause to scientists, philosophers, and scientific policy makers. The most controversial example is the Manhattan Project, where the research in nuclear physics necessary to design an atomic bomb was secretly funded by the American taxpayer, while the so-called Interim Committee, a handful of political, military, and technological leaders under a wartime administration, directed its construction and advocated dropping it on Japan.

The appropriate use of the scientific information from the federally funded technological advances of the Cold War remained a serious concern as the United States and the USSR wrestled for geopolitical dominance. The struggle included the Space Race and, in the end, spawned the Science Wars, an interminable debate between scientists and science critics about who should have the last word in setting the direction and constraints of science conducted with public funds (Fuller 2004). Scientists of the Cold War era largely refrained from criticizing the government's use of their research, in return for freedom of inquiry. This pact with the devil has deep historical roots dating at least to the Age of Reason, but NASA was born of the Cold War, and, since the collapse of the Soviet Union, it has suffered the slings and arrows of the Science Wars.

Science often gives rise to technological opportunity because it lets us think clearly about the unknown and to use that thinking to solve a problem. Progress flows more smoothly if trained scientific thinkers set the research priorities within boundaries set by informed citizens on the ethical constraints of the knowledge. This means a literate and principled society is necessary to decide whether new discoveries should be reduced to practice.

Space exploration is an opportunity both to think differently and to develop practical solutions to modern problems. The space program gave technological legs to unforeseen solutions to difficult problems. This is called "spinoff." The term implies that some problems can be solved only by a unique line of thinking, perhaps not directly related to the problem, that lays the groundwork for that solution. Spinoffs abound not just in the space program but throughout the history of science and technology. Chance does indeed favor the prepared mind.

Space science has produced many spinoffs, for instance, in aviation efficiency and safety, which we tend to take for granted, as well as in new technologies for monitoring and purifying air and water, satellite communications, the small size and high resolution of infrared cameras, and better methods of food preservation, like ready-to-eat meals for soldiers and refugees. In space medicine, the need to track the well-being of the astronaut remotely from Earth in the 1960s later gave rise to telemedicine: the transmission of a patient's heart rhythm from the ambulance to the emergency room or from pacemaker to the cardiologist, for instance. Other biomedical spinoffs have included better kidney dialysis machines, better hazardous-gas detectors, and greatly improved prosthetic limbs.

There are two other reasons that space exploration is important. First, space science gives us the capacity to detect certain problems before they harm us. We are all skeptical of forecasting, but we also take it for granted. Satellite tracking of massive weather systems, like blizzards and hurricanes, provides early warnings that save lives. Other satellites identify and track emerging problems before they become critical, such as the size of the ozone hole in the stratosphere or the rate of the melting of the Greenland ice sheet from climate change.

The acquisition of reliable scientific data on these world-changing physical trends is crucial to understanding them. The ozone layer over Antarctica has been monitored every year since the run-up to the International Geophysical Year in 1956 because it protects us against damaging ultraviolet rays. By the 1970s, an alarming expansion of the ozone hole was detected each spring, which led to the discovery of a major culprit, chlorofluorocarbons (CFCs). This led to an accord, the 1987 Montreal protocol, and to an action, a phased-in worldwide ban of CFCs in new products. Common sense tells us this is a success, even though no one ever formally proved that CFCs alone were responsible for the hole or that ultraviolet radiation passing through a colossal ozone hole would seriously increase our cataract and melanoma rates.

Climate change is altogether a different problem. We have failed to act because we cannot agree on the science, and without science, someone might as well just flip a coin when deciding what to do. Our planet is warming, but Mars is, too. Atmospheric CO_2 levels are rising, but CO_2 is neither a pollutant nor the only greenhouse gas; methane (CH_4) is twenty times more powerful. We know that we contribute to climate change but not how much. The climate models used by the Intergovernmental Panel on Climate Change

(IPCC) are under fire for predicting huge anthropogenic effects on sea level without considering that our climate resonates with or is synchronized to a set of natural frequencies in the Solar System (Scafetta 2012). We know that too much of a rise in the sea level is bad for our coastal infrastructure and commerce, but what about for the planet we live on? The only thing that makes sense here is more research.

Second, there is a rationale for pure research in fields like astrophysics, which strives to understand the universe. Each new astrophysical discovery, such as exoplanets, supermassive black holes, and the universe's accelerating expansion, changes our perception of the world and our relationship to it, including the Earth as a system and as a member of systems—the Earth-Moon pair, the system of inner planets, the Solar System, and the galaxy. For instance, the Moon stabilizes the Earth's tilt, which helps steady our climate, and changes in space weather can disrupt our communications and power grids.

The science of space exploration fits the historical flow of physical, biological, earth, and information sciences and the rapid technological pace of the twenty-first century. It interfaces with advances in genetics, neuroscience, computing, artificial intelligence, and robotics. The assimilation of new information and its dissemination into society transforms our ideas of how things work and what is possible, and these ideas are constantly evolving. Thus, human space exploration is an embodiment of modern scientific discovery. These important themes have shaped the worldview expressed in this book.

Our world is changing and so is our worldview, and human space exploration is a force for this change. There is no argument that it is not safer and cheaper to send robots rather than people to explore Mars or Titan or to search for offworld life. The tiny field of astrobiology has been fueled by the discovery both of an ever-widening array of earthly extremophiles and of thousands of new exoplanets. No matter how skeptical of life on other worlds, no one with any imagination can dismiss the possibility. The issue is oddly woven into our consciousness, even if the notion of finding alien "footprints" by peering through massive telescopes strains the credibility of all but the most sanguine.

Exobiology may not be recognizable, or it may be weird, but like those of physics and chemistry, the principles of biology may be universal. The right temperature and the right chemistry on the right planet might lead inexorably to life. The discovery of an Earth-like planet with signs of life would be stunning, but such an assertion would also be highly contentious. Just recall the furor in 1996 over Mars asteroid ALH84001, which was put directly under the microscope!

Extraterrestrial life is a needle in a haystack, and finding it probably *would* require a microscope; hence astrobiology has all but fallen off the radar. However, the telescope is telling us in which haystacks to search. We are such natural explorers and so deeply curious about our place in the universe that *space exploration is ultimately about us.* But we evolved here, and our biology practically forces us to stay here. Life is thus the purpose of space exploration as well as its specter. That is also a theme of this book.

Spaceflight is sometimes likened to the high adventures of the Age of Exploration, when tough explorers succeeded by painstaking preparation and by trial and error. They passed along their wisdom until technology caught up. This time-honored method breaks down in space, where life-sustaining technology is the heart of everything. If the technology fails, the mission fails, and people die. Today's technologies are adequate for low Earth orbit and for the Moon but not for Mars or beyond. Deep space calls for technologies that interface seamlessly on the edge of nothingness not just for months or years but, for all practical purposes, in perpetuity. This imperative is synoptic with human space exploration, and it too drives this book.

For the traditional explorer, the survival imperatives were water, food, and shelter. The modern explorer can visit a more deadly range of environments, for instance, underwater, where even air is lacking. However, the lack of air in space is a different animal all together, because it is so far from home and safety. Space cocoons us inside our technology, so we need to tinker with and perfect that technology in space. This was part of the rationale for the ISS; the closer to home we tinker, the safer we are. We even have our own splendid natural platform in near space, the Moon, despite some odd commotion about it being boring. Even if the Moon were boring, we have few other practical options. Asteroids aside, our searingly hot evil twin, Venus, lies in one direction, and in the other is Mars, tantalizing but farther away, with a wafer-thin atmosphere and the deep freeze of Antarctica.

What about that asteroid mission, for instance, to one of the Apollo group? This plan plays on subliminal paranoia over Earth-orbit-crossing asteroids, but must people go? Asteroids of this type are tiny (unless they are about to hit Earth) and can be explored (and nudged away) more safely and cheaply by robots. We can visit one, but we could not stay. Why substitute an asteroid for the Moon, and why do it now?

All nearby celestial objects share two problems: they are the wrong temperature, and they lack air. The Moon oscillates around 500° F and has essentially no atmosphere. These problems are exactly the same as those of low Earth orbit. Low Earth orbit has let us size up microgravity and troubleshoot advanced life-support systems, but it is not optimal for implementing other important technologies that we need in space. The Moon is essentially just overhead and has resources and intrinsic value in teaching us how to do this correctly. This is the best value in human exploration; therefore, returning to the Moon is an explicit agenda of this book.

Using the author's prerogative, the first chapter also lets some air out of the distracting "human versus robot" dichotomy in favor of integrated exploration led, of course, by robots. This is a pragmatic approach that optimizes a return on investment in space exploration and that has been steadily evolving for fifty years. Since the Apollo program, it has been clear that a robots-first orientation is the road to success. Probes will establish the physical parameters—the exact environment—as well as the level of scientific interest in distant worlds. This creates a "Goldilocks problem"—robots cover distance more safely and cheaply than people, so how do we prepare for both kinds of missions, unmanned and manned? There is no certain recipe, but the *proper timing* of each step is an important principle.

Since proper timing is inescapable, this book too is arranged on a timeline. Part 1 is a chronology of how we reached our current situation. Part 2 deals with contemporaneous plans for a home away from home, emphasizing new technology, robots, people, and the Moon. Part 3 explores our near-term prospects for the exploration of an ever-expanding final frontier. We will undeniably get to Mars, but just when no one can say for sure. I have no crystal ball for Mars or for deeper space exploration, but it is clear that we face bottlenecks that will require breakthroughs in science and technology in the decades to come.

Space is unconquerable in the traditional sense, and I will depend on hypothetical examples and analogies to explain why. For instance, the astute reader may have noticed that planetary scientists are developing a type of space biogeography comparable in some respects to the biogeography of remote places—isolated islands and mountaintops on Earth. This analogy helps us see the issues but is limited by not knowing whether life exists or can persist beyond the Earth. The analogy also incorporates two closely connected problems of scale—time and distance. For Mars and beyond, the time taken in travel will involve prolonged exposures to radiation and microgravity and conservation and replenishment of critical resources such as oxygen and water. These are two key benchmarks, but eventually the scarcity of all resources will preoccupy us in the vast, highly eccentric reaches of space. And outside the Earth's atmosphere and magnetic field, we must avoid the lethal radiation that rains down from the Sun and cosmos.

In the next few chapters, you'll see how temperature, atmosphere, water, food, microgravity, and radiation set the stage for the establishment of a meaningful human presence in space. But first, you may want to know how these problems evolved and have influenced the balance between men and machines.

PART 1
HINDSIGHT AND FORESIGHT

1. MEN AND MACHINES

In December 2003, America's commemoration of the hundredth anniversary of the Wright brothers' first heavier-than-air flight took place in Kitty Hawk, North Carolina. I had considered riding down to the Outer Banks from Durham to watch the event, but it turned out to be a cold, rainy day, and I wasn't too excited about seeing the actor-turned-scientologist John Travolta introduce President Bush. At Duke, the folks in our lab were hoping Mr. Bush would announce NASA's return to the Moon. I skipped it, nothing happened, and the Wright biplane replica didn't get off the ground.

The celebration of powered flight is not the only noteworthy anniversary of mankind's achievements above the Earth. The pressurized cabin, invented by the balloonist Auguste Piccard, had marked its seventy-fifth birthday, and *Sputnik* was forty-six. Yuri Gagarin's first orbital flight saw its fiftieth anniversary on April 12, 2011, and the *Apollo 11* lunar landing celebrated its fortieth anniversary on July 20, 2009. In 2010, the International Space Station (ISS) celebrated ten years of continuous habitation. These milestones passed quietly, but what a remarkable pace for such great achievement! In the twentieth century, our technology took us from being Earthbound to spending months at a time in space.

In 1972, Harrison Schmidt and Eugene Cernan left the Moon on *Apollo 17*, and no one has been there since. The Moon-Mars timetable had planned a return for 2020 (NASA 2004a, Congressional Budget Office 2004a, Brumfiel 2007), but there was no funding for it. A manned Mars landing planned for 2035 went up in smoke when Project Constellation became mired in controversy. A mission to Mars, despite rhetoric to the contrary, remains in the realm of science fiction.

A HOUSE DIVIDED

The United States and Russia have poured huge amounts of resources into their space programs for fifty years. Canada, Japan, and Europe have more recently entered the arena, but the costs continue to escalate. For a decade, students have heard me complain in my lectures about a "bottleneck" in human space exploration. In 2003, China threw its hat into the ring by launching a man into orbit, and, on paper, it has charted an ambitious space program. The Chinese have proposed, if they can afford it, landing astronauts on the Moon by 2025, and the former NASA administrator Michael Griffin noted that they might get there before the United States is able to return. Cost will keep things from escalating into another Space Race, but you can already see why we cannot get out of the spacefaring business altogether, lest we willingly cede our technological edge.

Americans blame NASA for an unimaginative and inefficient space program, and NASA bashers cite poor management, mountains of red tape, a lack of vision, and industry pork (*New Atlantis* 2005). This barking grew louder over the last few years of the shuttle program, although similar complaints can certainly be leveled at any large federal agency. I gently remind people that NASA invented America's space program; they put men on the Moon, built the ISS, and send sophisticated remote probes billions of miles to other planets with precision and accuracy—all under close public scrutiny.

NASA's large federal and civilian infrastructure makes use of a bit less than 1 percent of America's tax dollars. This budget has been disparaged as a work subsidy for America's rocket engineers and space scientists and as an inefficient expenditure of taxpayers' money. Yet NASA is neither a beneficiary of federal largesse nor responsible for stockholder-driven aerospace companies.

The agency has always operated in reaction to the pressures and constraints of a fickle bureaucracy and a politicized aerospace industry.

Although NASA is not the problem, there are problems at NASA. For decades, the agency was caught in a vicious cycle of trying to maintain in-house expertise in all areas of space research. This is a formula for trouble because something inevitably goes wrong. And when it does, the manager micromanages, the persnickety snuffle and stifle, and the innovator vanishes into the private sector. Even contract work is made tougher when the "all-stop" mentality takes over, with NASA experts scrutinizing every decision. These issues have been voiced within NASA and by outside contractors and laboratory directors for years. So has the need for less intrusive documentation and more constructive review policies. This was emphasized by the Columbia Accident Investigation Board (CAIB) in 2003, which faulted the managerial culture at NASA (CAIB 2003, NASA 2004b, and NASA 2007).

Most federally sponsored research in the United States is funded by grants issued to scientists in academia and private industry. The National Institutes of Health, for instance, funds biomedical research mainly through a large extramural program, and its infrastructure is supported by separate budgetary lines. The best projects are initiated by investigators, and the worst are mandated by Congress. By analogy, one could argue that America's space dollars should support NASA engineering and transportation infrastructure separately from NASA-funded, peer-reviewed science. Depoliticization would buffer NASA programs against partisan realignments and engage universities and corporations in long-range strategies to interface their scientific excellence with NASA's spaceflight expertise and infrastructure. This could insulate America's long-range investments in space from abandonment for the political reasons *du jour.*

Financial realities have forced NASA to outsource certain things, such as cargo delivery to the ISS, in order to maintain a pioneering program (Aldridge et al. 2004). In 2008, Congress approved payments to the Russian Federation their Soyuz spacecraft to support the ISS, and this outsourcing will continue until our new system is ready. In 2009, under a Commercial Orbital Transportation Services (COTS) program, NASA gave commercial contracts to Orbital Sciences and Space Exploration Technologies (SpaceX) to ferry cargo to the ISS. The SpaceX design, called the Dragon, is a reusable spacecraft consisting of a pressurized capsule and unpressurized trunk for Earth-to-LEO (low Earth orbit) transport of pressurized cargo, unpressurized cargo, or crew. This transition to privatization marked by the Dragon's first rendezvous with the ISS in May 2012 is a new beginning, but it is still expensive, and private companies expect to turn a profit.[1]

A decade of fiscal constraint created a proverbial house divided in the space exploration community. In 2006, NASA had proposed finishing the ISS by 2010 via sixteen shuttle flights, which meant flying a system five times a year that had flown only twice in the previous three years. To the frustration of its international partners, NASA cut $330 million from its biological and life-sciences programs in order to finish the ISS before retiring the shuttle and to jumpstart Project Constellation. The International Federation of Professional and Technical Engineers snorted, recommending instead scrapping the shuttle even earlier and moving on (Klotz 2006).

NASA stuck to its program, weathering the shuttle foam fiasco, completed the ISS in 2011, and retired the shuttle. Unfortunately, in the disagreement's aftermath, today we lack spaceflight capability. Until our space transportation system is ready for people, perhaps by 2020, we will pay the Russians or commercial ships to get crew and cargo onto the ISS in a trickle. This has obvious financial and strategic disadvantages.

This situation is the direct result of a flawed Moon-Mars initiative that allowed for minimum new capital and required the reallocation of already encumbered funds to expensive new technology. The fixed resources were split among the shuttle, the ISS, the new hardware, and the eight-hundred-pound gorilla in the room, a $4.5 billion science program.

The shuttle was thirty years old and obsolete, but designing and acquiring new hardware is very expensive and competes with the earth and planetary sciences programs, which too are crucial to U.S. space exploration's future viability. For any new mission to the Moon to be successful, the exact locations of recoverable water and oxygen-rich soils on the Moon are vital factors (Lawler 2007). Luckily, efforts to collect more data for a lunar exploration plan were not derailed, and in 2009 the Lunar Reconnaissance Orbiter (LRO) provided new information on lunar water and allayed the concerns that there might be none there at all.

The tense situation of 2006 and 2007 was followed in 2008 by massive cost overruns in the Mars Science Laboratory at the Jet Propulsion Laboratory. Over the years, under a quarter of NASA science missions have come in on schedule, and a quarter are well over budget, producing a domino effect (Lawler 2009). Cost overruns will never go away, but the large cash shortfalls in both science and spaceflight have produced an unworkable model. Indeed, NASA had recruited a highly talented science director, S. Allen Stern, who left in frustration after just one year.

During the 2009 economic recession, the situation became extremely bleak, flattening the NASA budget and ISS funding and causing decades of international investment to languish. The main ISS research laboratory, the 1.5-billion-dollar, 4.5-meter-diameter Columbus Module, was not installed until February 2008, after some twenty-five years in the making. This module, the European Space Agency's (ESA) main contribution to the ISS, had been planned for 1992, to coincide with the five-hundredth anniversary of the discovery of the New World. Equipped with ten telephone-booth-sized racks, it was designed for experiments in the life sciences, materials science, and fluid physics. It has a ten-year life expectancy, and the ISS crews, coached by ground researchers via video and data link, perform the experiments.

ESA touts its studies on microorganisms, cells in culture, and small plants and insects, as well as the European Physiology Module, which can measure the long-term effects of spaceflight on the body. This expensive infrastructure compares how organisms behave in microgravity versus Earth gravity. However, because the ISS is protected from most cosmic radiation by the Van Allen belts, which limit its detrimental effects on astronauts and other living things, radiation research on the ISS is not cutting edge.

The Columbus laboratory inadvertently generated controversy over another instrument, a seven-ton cosmic ray detector, the $1.5 billion Alpha Magnetic Spectrometer (AMS), designed to study the invisible universe. The AMS searches for the existence and distribution of dark matter and antimatter. The scientific reviewers found these issues innovative and important, but for a while, it looked like the AMS would never fly. In 2008, Congress finally authorized an extra shuttle mission to convey the instrument to the ISS. STS-134 carried the AMS up in April 2011, and it was duly installed on the ISS National Laboratory.

The issue of timeliness also affected the Japanese Experiment Module, Kibo, which was twenty years in production. Kibo includes a facility, a "terrace," that allows nonbiological studies to be conducted in open space, but such studies are still being conducted under the protective wings of the Van Allen belts.

These are examples of "science too little too late," resulting from the expense and time it takes to integrate experiments into a mission and to train neophyte mission specialists to conduct them. This waste of time and money really irritates scientists. Escalating costs along with federal budget cuts have endangered the ISS, which could be prematurely abandoned because of a lack of funding. The ISS would then be remembered mainly for the construction of the *station itself*, not for any of its scientific discoveries. This is a tangible legacy for NASA, but many people have hoped for something more.

NASA's forte is its superb engineers, and its expertise is building reliable space hardware. Only NASA has experience ensuring a seamless interface between space technology, people, and the environment. Historically, attempts at keeping the astronaut's environment as normal as possible has caused a lag in human space biology research. The logic of trying to maintain normalcy aboard a spacecraft fails because the astronaut's environment is not normal, thanks to microgravity (or partial gravity on the Moon), radiation confinement, and perpetual twilight. These factors have broad implications, but understanding them will require longitudinal human studies in space.

Microgravity has stumped science and medicine since it was recognized the 1950s, and some engineers want to solve the problem by making it disappear: using "artificial gravity," for example, with rotating spacecraft or personal centrifuges. But enormous spinning tops are hard to control, and on the Moon, it may be that partial gravity can stress bone and muscle enough to protect them from osteoporosis and atrophy, respectively. We also have a lot to learn about cosmic radiation, especially about physical and biological shielding beyond the Van Allen belts (Cucinotta et al. 2002).

The strategy of establishing orbiting stations that are easily supplied from Earth was the selling point of the ISS, and it holds for any future Moon expedition, as well. We will learn to use indigenous lunar resources while still being close enough to receive some support from Earth. Proximity to Earth reduces the consequences of making mistakes, which won't be the case when landing on a hurtling asteroid or on distant Mars.

ROBOT DAYS

As a rule, NASA's planetary sciences and robotic exploration programs exemplify how space research should be conducted by expert investigators. Robotic hardware costs a tenth of what manned hardware does, and it can yield truly spectacular planetary data leading to a better understanding of the origins of our Solar System and life on our planet. In the 1990s, this fact heightened the "man-in-space" problem, as America drifted away from the paradigm of the astronaut in the white suit and toward robotic hardware (Lawler 2002). Influential physical scientists argued for a strictly robotic paradigm because electronics and optics are cheaper and can reach farther than human exploration. The American Physical Society's rebuke of Moon-Mars pointed to the successes of probes such as the Mars rovers and the NASA-ESA *Cassini-Huygens* Saturn mission (APS 2004). The point is indisputable, but it is also not fair to compare apples and oranges. Both robotic and manned space exploration have their place.

The Mars rovers *Spirit* and *Opportunity* landed on the Red Planet in January 2004 and have returned to us reams of photographic and geochemical data about Mars. *Cassini-Huygens* completed a near-perfect seven-year mission to Saturn, collecting gigabytes of data about the planet, its rings, and its moons, especially Titan. In January 2005, *Huygens* set down on Titan, broadcasting data to Earth during its descent and for two hours on the surface.

These robotic missions far eclipse our meager human spaceflight capabilities, but robots have their limitations too, and failures are common. Probes are sent throughout the Solar System, even to asteroids and comets, but each is directed principally to one or two targets. Instrumentation and sampling sites are prearranged, and routes and routines are not easily changed across the vast expanses of space. A radio signal from Earth to Pluto, for instance, takes four hours to travel each way.

Robotic missions are successful when a probe hits the target and returns some data. For planetary objects, this involves a miniscule area, because the planets and their "little" moons are actually huge. Although considerably smaller than the Earth, since it lacks oceans, Mars has roughly the same amount of land area as our planet, and the largest asteroid, Ceres, has more land than Texas and Alaska combined. Clearly, astrophysical jargon like "tiny" and "close" is based on a different frame of reference than that of the traditional explorer.

The multiyear performance of the Mars rovers, which had been designed for ninety-day missions, is amazing and bears out the value of probes for planetary exploration. *Spirit* retired in May 2011 after a 4.8-mile trip, and *Opportunity* reached the rim of the Endeavor crater some thirteen miles from its landing site, remaining active into 2012. In their first year of operation, however, the rovers crawled about one hundred meters a day, which an astronaut on foot could cover in about two minutes. They confirmed a history of surface water but found no evidence of life. In the first Martian year (687 Earth days), the rovers traversed a golf-cart-wide path for about seven miles. This is like crawling forward on your hands and knees for two years in order to understand the whole of the Earth's geology. It makes sense if you've never seen dirt before and if the region is of particular interest, but for the Red Planet, this claim has worn a bit thin.

Remote sensor technology is improving at a phenomenal rate, yet every probe built and launched from Earth is obsolete by the time it arrives at its destination. This obsolescence reflects Moore's Law, which originally held that the number of transistors that can be placed in an integrated circuit doubles about every eighteen months. Figuratively speaking, electronics moves even faster than robotic spacecraft. As long as Moore's Law is in play, the obsolescence factor is overcome only by spacecraft velocity. Moreover, the complex instruments needed for these measurements are susceptible to failure as they drift for years through space. A one-way trip is the norm, and samples are rarely returned to Earth for study.

Case in point is the *New Horizons* spacecraft, the first mission to Pluto and the Kuiper Belt, which launched on January 19, 2006, on a ten-year, three-billion-mile mission to the outer Solar System. The Kuiper Belt, a distant region of ancient, icy bodies, is a fascinating and variegated feature of our Solar System. There are other Kuiper objects the size of Pluto—as well as the Kuiper Cliff, an area cleared out by some mysterious large object, perhaps the elusive Planet X, the Sasquatch of the Solar System. When *New Horizons* was launched, Pluto was defined as a planet, one with three moons; now it is considered a "dwarf planet" with at least four companions. Pluto lovers can take solace in the word *plutoed*—meaning "demoted"—the American Dialect Society's 2006 word of the year. But notwithstanding its newly plutoed status, Pluto; its barycentric companion Charon; two other tiny moons, Hydra and Nix, discovered with the Hubble (Schilling 2006); and the recently discovered P4 are interesting for many more reasons than when the mission began.

New Horizons is the size of a piano and weighs half a ton. So far from the Sun, solar power is not possible, and the spacecraft, like *Cassini-Huygens*, is powered by a radioisotope thermoelectric generator (RTG). The RTG contains plutonium, which its discoverer Glenn Seaborg named after the then ninth planet. The RTG contains roughly eleven kilograms (24 pounds) of plutonium dioxide packaged as ceramic pellets that generate heat from the isotope's radioactive decay. This heat is converted to electricity by solid-state thermocouples, initially providing about 240 watts of power. Over a decade, this falls to about 200 watts, roughly the same amount as a handheld hairdryer.

In 1821, a reliable apparatus to convert heat into electricity, the thermo-couple, was invented by Johann Seebeck, who connected dissimilar metals at different temperatures to each other. Thermocouples, having no moving parts, are ideal for use in space. In the RTG, the heat is generated by radioactive decay, so the isotope's half-life is important; for instance, the plutonium on *New Horizons*, though expensive, has a half-life of eighty-seven years. Available cheaper isotopes have half-lives that are too short for a ten-year mission. Construction and use of an RTG always triggers the concern and wrath of antinuclear groups, even though the Department of Energy had calculated the probability of *any* plutonium being released by a launch failure at one chance in 350. Compared with the actual catastrophic failure rate of the shuttle (five times higher), the risk of launching twenty-four pounds of plutonium into space to learn about the edge of the Solar System is negligible.

New Horizons was the fastest probe ever flown, whizzing past the Moon in nine hours, a three-day trip for Apollo astronauts. In early 2007, it flew to within 1.4 million miles of Jupiter, sending back photographs and data. *New Horizons* passed through Jupiter's powerful magnetosphere and found huge plasmoids, enriched in oxygen and sulfur, pulsing far down the magnetotail from the moon Io (Krupps 2007, McComas 2007). Jupiter's mighty gravitational field then hurled the craft toward Pluto at fifty thousand miles an hour, shortening the mission by three years.

In 2015, *New Horizons* will fly to within six thousand miles (9,600 kilometers) of Pluto and take photographs, record spectra, and collect physical data on the dwarf planet. For five months, it will have a clearer view of Pluto than the Hubble Telescope does. In 2006 dollars, the total cost of the mission was a mere $650 million. This is cheap for a potentially big payoff.

New Horizons says two things about human spaceflight. First, failsafe power is critical in space, and second, low velocity and interplanetary distance makes sending people unworkable. The longest anyone has spent in orbit is fourteen months, and no one could live in deep space for ten years on old technology. Until we invent high-speed interplanetary propulsion, the huge outer Solar System is beyond the reach of human exploration. A life in the great voids between planets is not for terrestrial humans, but *New Horizons* technology could make Earth–Moon journeys as convenient as a trip from Chicago to Tokyo on a jetliner.

Time and distance are part of the ebb and flow of the history of exploration. It takes time to collect the knowledge and experience needed to cross any frontier. In space, time has a dual role—the *lead* or *learning time* and the *proper timing*. Lead time and proper timing together determine when the progress actually occurs, as transportation and life-support technology are interleaved. In other words, the frontier settles out at a point determined by the balance between technological capability and travel time.

New technology is expensive, and only market demand and "fiddling time" brings the cost down. The Moon is in low in demand, and lunar missions need a long fiddling time. A moonbase will come later and be more expensive than most people realize, but a moonbase is the necessary step before a Mars base, and Mars is a necessary step before we reach the outer Solar System. Only then can we begin to think about nearby stars.

The huge distances of our Solar System are shown by the scale of the orbits of the eight planets. The inner system, including the Asteroid Belt, is only one-sixth the size of the outer system between Jupiter and Neptune, and the Kuiper Belt is larger still. Exploration of the outer system must rely on probes until fast propulsion, shielding, surplus power, and foolproof life support are in place. In theory, power, oxygen, and water can be generated from elemental hydrogen by fusion—"star" technology. This is certainly the future, but right now it remains the future: the world's state-of-the-art Tokomak fusion research reactor won't reach the power breakeven point until 2020 at the earliest.

SO FEW PLAUSIBLE OPTIONS

Our greatest experience in space has come from the extravehicular activity (EVA) needed to assemble the ISS. People built an infrastructure actually to live there. The ISS is proof-of-concept for a lunar base that would allow us to learn to live outside the protective Van Allen belts. The pursuit of international collaboration, cost sharing, and commercialization of the Moon has become hackneyed, but nothing else is sustainable. The Moon is the only platform large enough and close enough to build a community in space.

In the original Moon-Mars proposal, Project Constellation would have chewed up the original $104 billion, plus another $2 or $3 billion a year, and still not gotten us back to the Moon by 2020 (Shiga 2009). Washington gutted it, leaving only the Orion capsule, which can be adapted to various missions, including a future Mars Transit Vehicle. A 3.9-billion-dollar contract was awarded to Lockheed-Martin in 2006 to build the Orion.

Constellation needed two new rockets, the Ares I and Ares V, named so because of Ares' association with Mars, and new iterations of those launchers may have taken us there (NASA 2007). The Ares V would have launched the lunar *Altair* lander (or Lunar Surface Access Module, LSAM) and an Earth Departure Stage (EDS) similar to the old Apollo system but reusable and 2.5 times larger. *Altair* and EDS would have docked with the Orion astronauts for the Moon trip. This system was oriented around existing NASA engine technology, but cost and feasibility drew fire from dissenters who championed the military's traditional Atlas V system. The criticism had merit, and the frightening cost overruns made Constellation a Washington tar-baby. When the project was cancelled in 2011, NASA turned to the Russians for space transportation.

According to NASA, the Space Launch System (SLS), built at the Marshall Spaceflight Center, is a new type of heavy-lift capability for human exploration. It will also back up commercial and international transportation to the ISS. This is NASA's first go at developing a flexible, evolvable system designed to support either crew or cargo missions and be safe, affordable, and sustainable. The SLS uses liquid hydrogen/liquid oxygen fuel and existing Space Shuttle main engines, along with solid-fuel boosters for main propulsion. It will lift seventy metric tons (roughly 154,000 pounds) into orbit, including a crew vehicle (the Multi-Purpose Crew Vehicle, or MPCV) based on the Orion design. As the SLS evolves, the lower-stage solid boosters can be modernized and the upper stage adapted for a new rocket engine, the J-2X, to increase its lift capability to 130 metric tons. This would exceed the capability of the Saturn V that carried Apollo astronauts to the Moon.

The list of workable destinations in the twenty-first century for this new system is short. The possibilities, costs aside, were spelled out in 2011 by NASA's Human Space Exploration Framework Team (HEFT) as LEO (low Earth orbit), HEO (high Earth orbit), and NEO (near-Earth objects such as asteroids), but with little prioritization. They also emphasized near-lunar space (Lagrange points L1 or L2 or lunar orbit) and the lunar surface before attempts at the moons or surface of Mars. The destinations are illustrated in figure 1.1.

It is safe to say that the organization of this book reveals my own predilections. Apart from the ISS, logic points to near-lunar space (near L2) and the Moon. The Moon is the only place for testing infrastructure crucial for habitats, radiation protection, and indigenous-resource utilization. It has

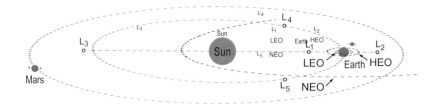

FIGURE 1.1. FEASIBLE DESTINATIONS FOR HUMAN SPACEFLIGHT IN THE TWENTY-FIRST CENTURY.
The diagram illustrates the solar orbits of the Earth-Moon (right) and Mars (left) systems. L_1 through L_5 are the Lagrange points in the Earth-Moon-Sun system, where the gravitational pull of two large masses (the Sun and the Earth-Moon) cancels out centripetal acceleration, enabling a third smaller mass to orbit at a constant distance from them (Earth-Moon Lagrange points are not shown). L_1 is home to the *Solar and Heliospheric Observatory* (*SOHO*) and other observatories, and L_2 is the destination for the James Webb Space Telescope, 1.5 million km (932,000 miles) from Earth. L_1 and L_2 may be used for future manned space observatories and supply depots. Other destinations are low Earth orbit (LEO), high Earth orbit (HEO), near Earth objects (NEO), lunar orbit, the Moon, the Martian moons, and Mars.

one-sixth of Earth's gravity, it is a prime location for astrophysics and to access the cryogenic near vacuum of space, and it is a goldmine of primordial geochemistry. Astronomers, already famous for putting their telescopes in inaccessible places, require little imagination for lunar-based astronomy. An ongoing lunar exploration program is critical to biophysical and planetary research and vice versa (Piantadosi 2006).

Misgivings about an NEO mission are prevalent, mainly because of the superabundance of nearby objects, our escalating robotic capabilities, and the overwhelming need for *surface technologies*, as we will see throughout this book. That epic visit to Mars will depend on these same technologies as well as on a substantial remote lift capacity. We are primed for the challenges, but a Mars mission simmers technologically, fiscally, and politically on the back burner for now.

2. A SPACE LEXICON

The space lexicon is famous for acronyms and buzzwords; some are quite useful, but others are peculiar or even confusing. As far as I can tell, the curious nonspecialist needs a simple working vocabulary that can handle three things: an idea of destinations and distances, a nomenclature for the technological and biological aspects of spaceflight, and a framework for the approaches that can limit risk and prevent damage to people who are planning to travel such distances.

Ours is the only habitable planet in our Solar System, so in space, we will be stuck living inside cocoons or artificial space *habitats*. We have learned to do this in the fringes of our upper atmosphere, in *low Earth orbit* (LEO). However, the ISS, which is roughly like living in a military cargo plane or a jetliner on a perpetual around-the-globe trip, represents the bare minimum in terms of habitat. We can see our next steps, which are limited to a couple of Lagrange points, the Moon, and some asteroids, with Mars further on down the list. One day we might visit a few moons in the outer Solar System or perhaps an *exoplanet* around a nearby star. But first, we must have habitats that, like our homes, are comfortable, efficient, safe, and durable.

Habitat development is a priority at NASA. When the crew arrives at its destination, they want to work and explore, not worry about their living space. Many experts find this a compelling reason to focus on the Moon and not an asteroid or Mars. The strategy of establishing habitats in advance of the inhabitants is also the prevailing opinion because it reduces the risk of catastrophic double failures. All missions have defined periods of risk, such as the launch, docking, and landing, and NASA requires human spaceflight systems "be designed so that no two human errors during operation or in-flight maintenance or a combination of one human error and one failure shall result in permanent disability or loss of life" (NASA 2003).

The ISS taught us that construction in space is challenging, analogous in some respects to assembling a submarine while underwater. Common sense dictates that submarines be built at the dock and then launched, so why can't the same hold true for spacecraft? Unfortunately, our planet is so massive that launching a submarine-sized spacecraft into orbit is beyond the capability of chemical rockets. Space offers no choice but to assemble large structures on site from prefabricated modules. The mass of each ISS module is substantial, and a heavy lift capacity is required for getting them off of Earth, but once in orbit, the components are essentially weightless and fairly easy to maneuver.

Advanced, lightweight habitat designs, including inflatables, are also being developed, and these could be deployed and assembled robotically in space or on the Moon. Such habitats are appealing because people do not have to be present to install them. The process would be the same whether the base is on the Moon or on Mars, but the steps are more expensive and complicated for the latter mainly because Mars is a thousand times farther away. The idea of a moonbase has had its detractors because even with the best technology, it won't be like home. Innovations in space transportation should allow us to go back and forth from Earth to Moon as we learn to live there and to cope with the naked cosmos. The moonbase can be supported from Earth as we learn to exploit lunar resources, and it will act as a bridge to Mars.

Space is harder to explore than other extreme environments, including the seabed, because of the unique engineering challenges it poses. Habitat design crosses disciplines, and habitat durability exemplifies the difference between *getting there* and *staying there*. This distinction will become increasingly important, and here I will focus mainly on the problems of staying there.

The five key problems are not the typical medical aspects of human space-flight one reads about in textbooks. Those problems include motion sickness,

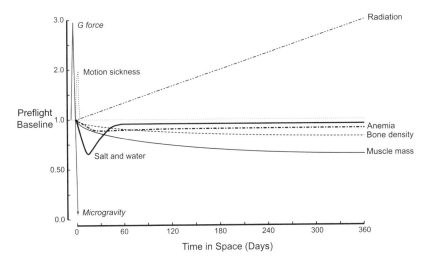

FIGURE 2.1. THE TYPICAL PHYSIOLOGICAL PROBLEMS OF HUMAN SPACEFLIGHT IN
MICROGRAVITY AND THEIR USUAL TIME COURSES.
Over a period of one year, the most important problems are radiation exposure and loss of bone
and muscle mass.

salt and water balance, anemia, radiation effects, and loss of bone and muscle.
A graph of how these problems develop over one year in space is shown in
figure 2.1. Of these, I will emphasize radiation and the effects of microgravity
on bone and muscle as well as three additional problems: life support, water,
and food.

First of all, however, staying anywhere requires energy. On Earth, our energy comes from a star, the Sun. In space, having a power source is just as
important, and the Sun again fills this role nicely for the inner Solar System,
but eventually, as we go farther out, it will become too dim to support us. As
distance from the Sun increases, the density of energy and other resources
diminishes, requiring conservation and reutilization.

On Earth, organisms settle into niches by evolving into their ecosystems.
These ecosystems are finely tuned, and our ability to disrupt them on a
global scale is familiar to us all. We may have put ourselves, the inhabitants
of "Spaceship Earth," in a precarious position for disturbing nature, but the
idea that we can expand our civilization and increase our population by using the resources of the rest of the Solar System is naïve (Boulding 1966).
This brings us to our first problem. You may have wondered how natural

ecosystems settle into quasi-steady states and act "just right." We simply don't know, and experiments with closed ecosystems on Earth, such as Biosphere 2, have been unsuccessful.

Reliable *life-support systems* supply a clean, breathable atmosphere of an appropriate composition and temperature. These systems really consist of multiple subsystems, integrated, backed up, and operating continually. Some engineers refer to this as "mature technology," which is true for low Earth orbit, because the life-support systems are supported from Earth. It is an altogether different matter to support such technology on the Moon and even more difficult on Mars, where you cannot call your NASA HVAC man to ride up on the next *Soyuz* to fix things.

Fundamentally, life-support systems are *nonregenerative* (open) or *regenerative* (closed). In regenerative systems, energy is spent to recycle used resources into some new resources, saving on resupply costs. Ordinarily, it is less expensive to repair your old car than to buy a new one. A closed system reuses everything by constantly recycling resources, using energy but no new stock and leaving no permanent waste. Perfectly closed systems are not allowable in thermodynamics, and shipments of cargo or harvesting of resources will be required to replace the resources that are irrevocably lost. NASA calls the latter *in situ resource utilization*, or ISRU, which we'll examine later with respect to oxygen and water.

Second, water and food are essential, and they are coupled to the removal of waste and the recycling of resources. These processes will eventually be integrated with atmosphere control into a controlled ecological life-support system. In essence, this would replicate the life cycle of the Earth and is the Holy Grail of space exploration.

Next is cosmic radiation, which requires monitoring and shielding, and then microgravity, which is tolerable for a short while but soon spells trouble for bone and muscle. Here, the optimal physical and pharmacological countermeasures must be defined. Finally, our long-term fitness for space is unknown because there are subtle issues with hormones, bone marrow, the immune system, and biological rhythms, as we will see in later chapters.

THE THIN BLUE LINE

Our atmosphere is much more than a breathable blanket of air; it, along with our magnetic field and our oceans, is our protection against solar and cosmic radiation. An *atmosphere* is any envelope of gas surrounding a planet or contained within an artificial environment such as a pressurized airliner, submarine, or spacecraft. Science fiction writers often pretend that we can be moved like tropical fish in a plastic bag of water from one aquarium to another, but it's more complicated than that: we have tight constraints for barometric pressure, temperature, oxygen, carbon dioxide, and pollution. A clean, breathable atmosphere at the right pressure—a stable bubble—is required of every manned spacecraft.

The Earth's atmosphere, the thin blue line as viewed from space, is the column of air of uniform composition bound to the planet by gravity. It is densest at sea level and gets thinner, or more rarified, with altitude. The mass of the entire air column exerts a pressure at sea level of 14.7 pounds per square inch (psi), 760 mm Hg, or 1,013 millibars (mbar).

Our atmosphere is many miles thick, and it is divided into five layers with different thermal characteristics. However, we can live unaided only at the bottom half of the bottom layer, the *troposphere*, which is about 30,000 feet (9,000 m) thick at the poles and 56,000 feet (16,800 m) thick at the equator. In fact, half of the mass of the entire atmosphere is found in the first 18,000 feet (5,400 m) from sea level up.

The mixture of atmospheric gases called *air* contains approximately 20.9% O_2, 78% nitrogen (N_2), 0.9% argon (Ar), 0.04% carbon dioxide (CO_2), variable amounts of water vapor, and traces of many other gases. These standard definitions of atmosphere and air are used throughout the book. A "standard atmosphere" is defined by an idealized profile of a steady-state atmosphere at moderate solar activity encompassing a group of physical parameters including gas temperature, pressure, density, and viscosity (Johnson 2002).

Artificial atmospheres may or may not contain the precise proportions of gases that constitute air, but if we are to breathe them, they must contain O_2. Closed atmospheres are prone to trace contaminants, some of which can damage the lungs. On Earth, natural processes, including atmospheric chemistry and precipitation, help clear pollutants from the air, especially soluble gases and airborne particulates. Solar and cosmic radiation also help rid the air of contaminants, but few, if any, pollutants dissipate into space.

The importance of our atmosphere is immediately apparent to anyone who flies on a commercial jetliner. As you peer out the cabin window into thin air, the ear popping and changing temperature are little reminders that outside pressure is changing but that the cabin is keeping you safe from the dangers of high altitude. A spacecraft cabin is the same, but with no

appreciable atmosphere outside. Other forces, such as acceleration, temperature, radiation, and microgravity, also factor into spacecraft environments
and affect the range of tolerance.

Our ability to tolerate such stresses varies, and not because of physics, which
is set quantitatively, but because individuals respond differently to physical
stress. These differences reflect our physiological abilities to adapt and lead to
differences in *tolerance* in people in the same environment. For example, only
one in four climbers who set out for the summit of Mt. Everest reach it, even
with the correct motivation, fitness, climbing experience, and acclimatization.
It is simply too far beyond the tolerance limits of some people.

The importance of adaptation is seen by comparing the aircraft and Everest examples above. A normal pretzel-eating business-class passenger would
pass out in two minutes if cabin pressure is lost at 29,000 feet, while certain
Sherpa can ascend Everest without supplemental O_2. This variability engenders various opinions and some disagreement among physiologists about
how long an individual can tolerate being close to such physical limits. To
understand these differences, we need the science of *limit physiology*.

THE SCIENCE OF LIMITS

Survival limits are part of a specialized field of medicine called *biomedicine*—
the study of the *human body and survival in physically stressful environments
and within protective alterations to that environment.* Biomedicine tells us how
we perform under stressful conditions. All medical research is essentially biomedical, but I will limit my discussion to processes that maintain biological
stability, or *homeostasis*, as our surroundings change. This definition sets environmental medicine apart from disease (*pathology*). Sharp lines between environment and disease are difficult to draw, but often the distinction is moot.

For instance, when you are cold, you shiver, and the blood vessels in your
skin constrict. But the cold also promotes coagulation and predisposes to the
formation of blood clots, which increase the risk of stroke and heart attack,
especially in older people. In other words, sufficiently intense or prolonged
physical stresses can cause disease. Moreover, specific environmental factors
may permanently alter one's risk for diseases like cancer and asthma.

Biomedical research is also not restricted to people but includes our evolutionary co-players, especially microbes. It is easy to see that microbes might
become less important to us in space, but we have coexisted for so long that

we may not be easily separable. Another staple is plants, which perform oxygenic photosynthesis and provide nutrients. Yet even with synthetic nutritional substitutes, no one knows for sure how we would fare without plants. Special consideration thus attends microbes and plants in long-term human activities in space.

The term *stressful* also has a physiological definition. It is used to describe any environment that places an excessive demand on homeostasis. This is easy to grasp in extreme conditions but hard if the environment seems normal. As long as you feel well, crossing the line can be hard to tell. For instance, it is not particularly stressful to walk through the arid desert during the early morning or even continue into the heat of the day, if you have a hat and water. However, within a few hours of draining your canteen, the desert becomes quite stressful. On the other hand, some unhealthy environments never seem stressful because there are no imminent effects, for example, in an overpopulated, polluted, drug-ridden metropolis.

Stressful environments by definition affect our physiology—our bodies respond physiologically to avoid putting life and limb at risk. If it is too hot, we sweat; too dry, we concentrate the urine; too cold, blood vessels constrict in the skin; and too high, our breathing increases (hyperventilation). These responses protect what the nineteenth-century French scientist Claude Bernard called the *milieu interieur* and the American physiologist Walter Cannon termed *homeostasis* (Cannon 1932). These responses protect us in two ways: the immediate physiological changes reduce stress on our system, and we can sense those changes and choose to change our behavior to avoid or further diminish those stresses.

A change in environment that invokes a response that subsides after the end of the exposure is called a *stressor* or *adaptagent*. When the stressor exceeds some threshold, the response is a *strain*. A constant strain may initiate *compensation*, which may allow a gradual decrease in the intensity of the body's responses. This *tolerance or habituation* is a hallmark of *adaptation*. In other words, a prolonged but tolerable strain leads to *acclimation* or *acclimatization*, which are reversible adaptations (Fregly 1996). The inability to compensate means that the stress is over the *limit* and that homeostasis will be lost. This reduces *survival time*. The famous "fight-or-flight" response is an acute frontline defense that includes the release of adrenaline and other hormones that increase blood pressure and heart rate. Although energizing, it can only be used temporarily, for escape, and cellular compensation must come into play to avoid injury or death.

These principles are all nicely illustrated by our responses to high altitude. As the altitude increases, the air becomes thinner as the barometric pressure (P_b) falls, but the percentage of O_2 in the air (the fractional O_2 concentration) remains constant at 20.9%. That is, the total number of O_2 molecules in any volume of air decreases with P_b. This air has a low O_2 content, a problem called hypoxia.

At about the altitude of Denver (5,280 feet), hypoxia causes us to breathe more deeply and rapidly, or *hyperventilate*. This keeps the O_2 content in our lungs up by lowering the level of CO_2. The reciprocity between the pressures of O_2 and CO_2 in the lung is an effect of Dalton's law of partial pressure — each gas in a mixture of gases (like air) exerts a partial pressure, and the sum of the pressures is the total pressure. As we go higher, the lungs receive less O_2, so we breathe harder to lower the lung's CO_2 level to make room for more O_2.

High altitudes are also cold, which invokes other responses to avoid *hypothermia*. The skin's blood vessels constrict to reduce heat loss, and our muscles shiver to generate extra heat. Shivering utilizes more O_2 and causes a further increase in ventilation. High altitude thus stimulates breathing by two mechanisms, hypoxia and cold. This exemplifies the fact that stressful environments often contain more than one stressor.

If the body responds immediately, the response is an *accommodation*, while a slower response, for example over days, is an *adaptation*. If one stressor is involved, the body *acclimates*, and if multiple stressors are involved, it *acclimatizes*. All adaptations have intrinsic rates, and the intensity, duration, and sequence of responses depend on the intensity of the stimulus, the body's defensive hierarchy, and the physical reserves.

Once a stress is removed, adaptive responses also dissipate at different rates. The reversal is usually complete, and unless an injury has been incurred, things return to normal. This reversibility also usually pertains to structures, such as the increased number of circulating red blood cells at high altitude or salt conservation in the heat. These principles of biological adaptation are not unique to us; they are shared throughout the vertebrate animal kingdom.

The stress responses depicted so far are transient and reversible, but permanent genetic adaptations also occur. These affect the lives of individuals, but they will affect community survival only if reproductive fitness is altered. Humans have managed to adapt in many extreme environments, but high altitude is perhaps the most difficult. Altitude is also an evolutionary challenge because, among other things, hypoxia impairs the ability of a mother to carry a fetus to term. This biology is fascinating in its own right, but it also teaches an important lesson about what might happen genetically to people in space.

GENES AND ADAPTATION

The highest human homes are at elevations above 4,000 meters (13,333 feet) on the high plateaus of the Himalayas and the Andes. The Tibetan plateau appears to have been settled about 24,000 years ago, and the Andean Altiplano about 11,000 years ago. These two groups of people both live where the air has only 60 percent of the O_2 at sea level, but they have followed different biological paths to life at high altitude.

The Tibetans have normal concentrations of the major O_2-carrying protein, hemoglobin, in their blood, while the Andeans have high hemoglobin levels, similar to lowlanders who acclimatize to high altitude. The indigenous women of both populations bear children at elevations above 12,000 feet despite the combined physiological stresses of pregnancy and hypoxia. There are, however, several problems.

Andean mothers develop the prenatal syndrome of preeclampsia more often than do lowland mothers, and the fetus develops more slowly because of the growth-inhibiting effects of hypoxia. Infant birth weights are generally about one-tenth below normal for gestational age, but they are higher for infants of native Andean mothers than mothers of European descent who live at the same high altitudes. Native Andean mothers have adapted mainly through the ability to increase the blood supply to the placenta.

Tibetans do not get mountain sickness or show maladaptive increases in blood pressure in the right side of the heart or the pulmonary circulation at high altitudes. During exercise, their muscles take up the same amount of O_2 as do those of people who live at sea level. This means Tibetans have found a unique way of overcoming hypoxia, the major limitation of the high life.

A few years ago, a team of anthropologists from Case Western Reserve University found that Tibetans compensate for low O_2 levels by doubling the rate of blood flow to the forearm muscles (Beall 2010). Presumably, this response also reflects the conditions in the major internal organs. The team found that Tibetans, compared with residents at sea level, had tenfold higher nitric oxide (NO) levels in their blood. Nitric oxide is a natural dilator of blood vessels and allows more blood to be delivered to the tissue capillaries.

This large amount of NO seems to have no ill effects, and the Tibetans are physiologically normal. However, it is not clear how they arrived at this state or whether it would eventually develop in lowlanders who permanently migrated to high altitude. Asked to comment, my innovative colleague Jonathan Stamler said it best: "These remarkable data in Tibetans provide a beautiful demonstration of how Nature has evidently exploited nitric oxide to offset the effects of high altitude."

Genetic studies have also determined that Tibetans have actually lived at high altitude long enough to show heritable changes. Only a handful of the twenty-five thousand human genes seem to have played a role in their adaptation, but the exact nature of the adaptation biochemically and physiologically has remained elusive. There are a few rather remarkable observations.

The first was the finding of a gene for which the inheritance of only one of its alleles is sufficient to confer a higher O_2 affinity to the hemoglobin of children of mothers whose families have lived at high altitude for many generations. Genes are made up of two alternative alleles carrying essentially the same information and occupying the same position in homologous chromosomes: one from each parent. The allele of interest helps the red blood cell take up O_2 as it pass through the lungs, and it offers a survival advantage to children who inherit it.

Another study found a genetic variant in Tibetans that seems to help them live comfortably at high altitude without having to increase the blood hemoglobin content (Xi et al. 2010). This variant is a point mutation, or SNP, in a gene called *EPAS1*. SNP is an acronym for single nucleotide polymorphism, the substitution of a single DNA base in the genetic code, and SNPs can significantly alter the function of the protein encoded for by the gene. *EPAS1* encodes a transcription factor, a protein that promotes the expression of one or more other genes. The EPAS1 transcription factor regulates genes that are involved mainly in the response to changes in the amount of O_2 available to the body.

This genetic finding created an intriguing puzzle for physiologists. Almost 90 percent of high-altitude Tibetan people carry the *EPAS1* SNP, but it is found in only 9 percent of the closely related Han Chinese, who live near sea level. Yet the two groups show no differences in blood hemoglobin or O_2 carriage. The investigators who discovered this phenomenon speculated that carriers of the "Tibetan allele" have an unusual way of maintaining the level of O_2 in tissues at high altitude without needing to increase the hemoglobin levels.

EPAS1 is a hypoxia-inducible factor, or HIF, a protein made in respond to hypoxia. HIF proteins regulate the expression of genes for glucose breakdown, hemoglobin production, and the density of small blood vessels and capillaries in the tissues, and their levels in the cell nucleus increase during persistent hypoxia. EPAS1 is found mainly in the endothelial cells lining the inside of blood vessels in the heart, lungs, and placenta and is implicated in the growth and development of new blood vessels. This implies that Tibetans have an adaptation that increases the blood supply to tissues that rely most heavily on O_2 and that EPAS1, through some unidentified relationship to NO, introduces a change in the structure and/or function of blood vessels.

The Tibetans have also undergone the fastest evolutionary change ever chronicled in humans. According to statistical models, the difference in the prevalence of *EPAS1* between the Tibetans and the Han compared with a control population in Denmark is such that the two Asian groups seem to have diverged from each other only 2,750 years ago. This is quite rapid, yet the Tibetans had been living on the plateau for roughly one thousand generations before the *EPAS1* mutation emerged.

What are the implications of such natural selection for space travel? Permanent adaptations in space will occur only when people live there permanently, so it is all but impossible to predict what will happen. This has not stopped people from speculating, of course, some thoughtfully and some wildly. I prefer the thoughtful approach, but there is a definite implication that given enough time, the capacity for human adaptation may be nearly inexhaustible, provided the leaps from one evolutionary lily pad to another are not too far apart.

Spacecraft environments, for obvious reasons, must remain under our control, but there may be enough wiggle room that you could reasonably expect spacefarers on a multigenerational voyage to arrive at a distant planet in a rather different biological state than their ancestors who left Earth. A gradual decrease in cabin O_2 concentration, for instance, might ultimately lead to a Tibetan-like resistance to hypoxia in the crew. This could be useful, because it would decrease the need for O_2 recycling on the ship. Of course, the Tibetans could decide that they already have an edge as space explorers and outcompete the Han for all the taikonaut slots! Similarly, over time, the resistance of the crew to cosmic radiation could be augmented.

Agents such as cosmic radiation and toxic chemicals can lead to adaptive or maladaptive (injury) responses. Composite responses may be confusing, and subsequent adaptations can be difficult to tell from maladaptive effects of injury or disease. Some physical changes, like genetic mutations, may be permanent but go undetected for years. The classical example is radiation exposure, which produces cancer-causing mutations as well as mutations in germ cells passed along to progeny. If these mutations become widely distributed and lead to decreases in fecundity, they are maladaptive. Other nongenetic, or *epigenetic* changes, to our traits are not transmitted in the genetic code given to your children but are instead temporary, transmissible chemical modifications to the nuclear proteins that help regulate gene expression.

Random mutations may profoundly affect small, isolated populations. Mutations have positive, negative, or neutral implications, where *positive* means the mutation imparts a particular trait with a reproductive advantage. That trait might give an offspring a greater chance of surviving a first winter or let it move into a less competitive niche. A long-term change in the environment may also allow a trait that originally had a survival advantage to be deselected at random. If neutral, the trait is not missed, but if the original stress is reintroduced, offspring who lack the trait may be in trouble. If loss of a trait becomes *fixed* (absent in all individuals of reproductive age), reintroduction of the stress may have dire consequences. In a small population, the fixation of such a trait may ultimately drive that population to extinction, a potential concern in space.

SOME ASTRONOMICAL CONCEPTS

Beyond Mars, distances are disproportionately high compared with spacecraft velocities, and as time in space increases, our biological problems get harder to solve. Earth to Jupiter is the equivalent of eight Earth–Mars trips. There may be a waypoint between Mars and Jupiter at Ceres, but Ceres is cold beyond belief. Saturn's Titan is roughly twenty, Uranus forty, and Neptune sixty Mars trips. A trip across the asteroid belt by conventional rocket is prohibitive: it takes too much fuel. Years of monotonous confinement in twilight, weightlessness, and radiation exposure, punctuated by the threat of cosmic storms, meteorites, and atmosphere degradation, would wear you out. Even with advanced propulsion systems, such missions may be so ardu-

ous that future astronauts would become the grim risk takers of science fiction who live fast and die young.

I have implied that *population ecology*, especially for endangered populations, may one day apply in space. The biogeography of islands that formed
the basis for the thinking of Darwin, Wallace, and other evolutionary pioneers may also predict our dispersal in space. This is illustrated nicely by an
analogy with space travel, which will also provide the opportunity to introduce painlessly some astrophysical terms into our discussion.

When large distances are encountered in this book, I have used astronomical units (AU) inside the Solar System and light-years or parsecs outside
the Solar System. One AU, traditionally set by the distance between the Sun
and the Earth on its axis (~93,000,000 miles, or 149,598,000 km), is defined
as the distance from the center of the Sun that gives a circular orbit an orbital
period of one Gaussian year (365.2569 days).

The speed of light, c, is 186,000 miles per second (~300,000 km/second).
A light-year (Ly), the distance light travels in a year, is about six trillion miles.
The Earth is eight light-minutes and Neptune four light-hours from the Sun.
The Sun's nearest neighbor star is 4.3 light-years away.

For distance, astronomers prefer the *parsec* (pc), shorthand for *parallax
second*, a trigonometric measurement of an object in the sky, like a star. This
requires observing it from two spots a known distance apart, for instance
when the Earth is at opposite sides of its orbit. *Parallax* is thus the apparent
motion of a fixed object because of a change in the observer's position.

Stare for a few seconds at one spot on this page, close one eye, and then
switch eyes. The page "jumps" horizontally relative to the background. This
is parallax, and it is inversely proportional to distance. Closer objects have
greater parallax than distant ones, and because the parallax of celestial objects
is tiny, precise measurements are needed to estimate their distances from
Earth. The parsec takes the 360 degrees of a circle and divides them into
angular, or arc, measurements. Each degree is has sixty minutes, and each
minute sixty seconds, or arc-seconds. One parallax-second is 1/3,600 of one
degree of arc, and thus one parsec is 3.26 light-years, or 206,265 AU.

Distances in light-years are set by observations at different points in times;
an object is fixed at the location where it was when the light that we are
seeing from it now left it (e.g., for the star Gl581, 20.4 years ago). The use
of light-years is reasonable for nearby stars but less so for distant objects
like other galaxies. Let's plan a hypothetical trip to the "nearby" class M star

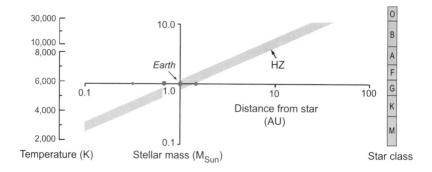

FIGURE 2.2. THE THEORETICAL HABITABLE ZONE (HZ) OF A STAR.
The HZ has a temperature at which liquid water can exist on the surface of a planet. It is a function of stellar mass and distance from the star. The Sun and the inner planets of our Solar System are shown on the X-axis. The Earth is located at the intersection of the axes. The dots on the X-axis to the left of Earth are Mercury and Venus, and to right is Mars. The Y-axis key indicates the types of stars. Both axes are on a log scale.

Gliese 581 (Gl581). This red dwarf is 20.4 light-years (6.26 parsecs) away in the constellation Libra (Udry 2007). If our ship travels at 1 percent of c, the voyage would take 2,040 years, or about one hundred human generations. Red dwarfs are long-lived stars, and thus we expect the star to still be there when we show up.

Gl581 has a planetary system. Two of its planets—the so-called super-Earths Gl581c and Gl581d—have attracted attention since their discovery (Udry 2007). Gl581c has a mass of about five times the Earth—at the time of its discovery it was the most Earth-like of all exoplanets. Gl581c orbits its star every thirteen days, at a distance of 0.073 astronomical units, which is only 6.8 million miles (10.9 million km). Given the luminosity of Gl581 (1.3 percent of the Sun's), the planet's surface temperature is probably between −3°C and +40°C. This is within the habitable zone (HZ) of the star, meaning that liquid water may exist on the surface, but its true temperature will depend on the atmosphere and the strength of the greenhouse effect. Gl581d is more massive, 7.1 Earths, but orbits at a more comfortable 0.25 AU, with a period of 83.6 days, placing it at the outer edge of the HZ. The width of the HZ varies greatly among types of stars, as shown in figure 2.2.

There may be up to six planets circling Gl581, although the unconfirmed Gl581g and Gl581f would also be closer than Mercury is to our Sun. All these planets are tidally locked—one side faces the star in perpetuity—and each has a cold and a hot hemisphere. And these high planetary masses mean a greater gravity than we are used to, but let's posit a breathable atmosphere for Gl581d, a smaller iron core, and a thick arctic mantle, making it less dense than Earth and giving it a surface gravity of 2 g. A 75 kg (165 lbs) astronaut would thus weigh 150 kg (330 lbs).

Suppose the voyage is made in weightlessness and that some of the crew carries a recessive gene for a disease called *dysautonomia*, in which people with two copies (alleles) develop severe orthostatic hypotension. These people faint from low blood pressure when they stand up (this is orthostatic hypotension). Carriers of one (heterozygotes) or two copies of the gene (homozygotes) would have no problem in zero gravity. They notice nothing; the trait is silent and undetectable without genetic typing. If this trait, for some reason, offered a reproductive advantage in microgravity, it could for reasons that don't matter here become fixed in all children born on the ship after forty or so generations. After one hundred generations, the ship arrives in orbit around the planet. If the ship's crew does not know that a genetic change has altered their fitness for gravity, a landing would be calamitous, because no one would be able to stand up on the planet (it has a gravity of twice that of Earth)—nor tolerate the g-force to return to space.

Orthostatic hypotension is an actual problem commonly seen when astronauts return from even a few days of weightlessness. Weightlessness has cardiovascular effects that cause fluid to shift to the face and upper body. This causes the cardiac stretch reflexes to act as though the circulation is overloaded, and the kidneys in response excrete "excess" salt and water (figure 2.3). This *diuresis* is familiar to every kid in a swimming pool, especially in cold water, but most astronauts still develop the "puffy-face bird-leg syndrome" in space.

Here is another example. Air conditioning offers relief from the summer heat, but most of us can still acclimate to hot weather. Physiologically, however, old folks adapt less well to heat, and if the air conditioning in a nursing home breaks down during a heat wave, the residents start to die. This happened during the European heat wave of August 2003: 14,000 people unexpectedly died in France over a ten-day period. The deaths were mainly among those seventy-five years old and older, and two-thirds were in hospitals, nursing facilities, and retirement homes that were not air conditioned. Infants and small children are also very susceptible to heat.

Behavioral adaptation comprises actions that avoid or alter an environment to prevent stress. We manipulate our environments to make them "better," and these "improvements" are good if we retain the ability to live in

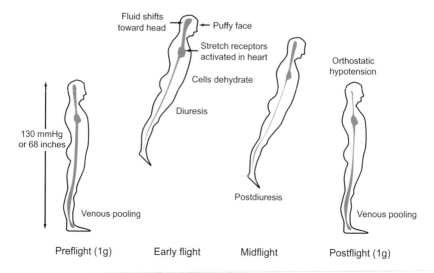

FIGURE 2.3. CHANGES IN BLOOD VOLUME DURING AND AFTER SPACEFLIGHT IN MICROGRAVITY.
The volume of blood plasma decreases in microgravity in response to the movement of blood into the heart and great vessels. This centralization of blood activates the body's baroreceptor (pressure) reflexes and stretch receptors, which causes diuresis and may lead to cellular dehydration. This results in orthostatic hypotension when the astronaut leaves microgravity and returns to Earth.

the native environment. This means that space exploration is not just about technology; it is also about biology.

Exploration is physically toughest when extreme conditions and vast distances are involved, and space exploration exemplifies both. Not surprisingly, the slow rate of space exploration is comparable to that of early continental or undersea exploration. Yet we are making progress, and no one doubts that our descendants will marvel at our having gone to the Moon on something as archaic as *Apollo 11*, just as we do at the *Santa Maria* crossing the Atlantic.

THE TORTOISE AND THE HARE

The hare may have won the race to the Moon, but today the tortoise has the advantage. Slow and steady goes a lot farther in space than haste and waste. All manned space missions share key elements, but if the goal is to stay and work in space, an asteroid mission is an expensive detour; a small moonbase is the steady, logical next step. The work on the ISS has been focused on understanding the problems of exploring the Moon and later Mars.

The stepwise logic in space exploration is the rationale for part 2 of the book, and those chapters cover the specific issues that must be settled before going to Mars. The architects of the NASA lunar prospectus hold the high ground here over those who fret over unnecessary delays in interplanetary travel. The Moon is in deep space but also practically next door; the more we use it, the better prepared we will be for Mars. A Mars mission requires landing on and leaving from another sizeable planet, not a tiny asteroid; it is very far away and offers no leeway for major technological failures. A race to Mars, even if it didn't cost trillions of dollars, increasingly seems like a bad idea.

Mars first captured our imaginations through the *canali* drawings of Giovanni Schiaparelli (1835–1910). These were misinterpreted as deliberately constructed canals, sounding the alien alarm. Modern data on terrain and atmosphere make it clear that Mars will not be easier to explore than the Moon. People will probably visit Mars in the second half of the twenty-first century and perhaps even stay in the following century. This may seem slow even for the tortoise; however, the appeal of the Red Planet is slowly being eroded by robotic explorations that have exposed its bleakness. Based on a sustainable rate of technological investment, a manned Mars mission with a high probability of success is fifty years away. Of course, this is the same estimate that was given after Apollo.

The other planets of the inner system, Venus and Mercury, offer different quandaries; both are hot, and Mercury is bombarded by solar radiation. The images of living too close to the Sun and rampant climate change on our own planet are enough to be hesitant about either. Our evil twin, Venus, orbits the Sun at 67 million compared with our 93 million miles (we are 1.0 and Venus is 0.72 AU). Venus is on the edge of the HZ for the Sun, and its surface temperature would normally be higher than Earth's by about 90°F (50°C). Liquid water might have once existed there, but a CO_2 atmosphere a hundred times as dense as ours has made it a runaway greenhouse, and its average surface temperature is 872°F (464°C). Venusian life likely would be similar to the tubeworms living near a hydrothermal vent 3,300 feet beneath the sea. Still, exobiologists speculate about life evolving on Venus, since it was the warmer incubator throughout prebiotic history.

Mercury is farther away than Mars but orbits the Sun at just 36 million miles (0.38 AU). Mercury is all but tidally locked to the sun; it has just three days every two Mercurial years. The hemisphere facing the Sun broils at 878°F (467°C), while the other freezes at −300°F (−183°C). It has no atmosphere, and it is a dense onion of iron and silicate as hard as a frying pan (5.43 g/cm³), although it may once have had additional, less dense outer layers that were baked or blown away. The core is partly molten, with depth and spin, and the planet has a weak magnetic field, about 1 percent of that of Earth (Margot 2007). It may contain an alloying element, like sulfur, which affects its melting temperature and may have allowed the core to remain a liquid for billions of years.

It would be conceivable to land on Mercury, in shadowed craters near the poles, and some of the deeper ones might contain ice. A peek up at the Sun would reveal an orb three times as large as it is on Earth. Mercury's proximity to Sol makes it unlikely that we would spend much time there, and the idea of being caught in a solar storm in transit is a terrifying prospect indeed.

In the outer Solar System, there is no hare, only the tortoise. The science required to travel there requires surmounting three formidable obstacles: distance, cold, and cosmic radiation. The cold reflects a falling energy density or falling energy per unit-area, elegantly demonstrated by the decline in temperature as the Sun's heat spreads out over an enlarging sphere. Since energy falls by the inverse square law (with the square of the distance), spacecraft will require more than solar power to travel beyond Jupiter. Distance aside, we are also constrained by radiation and gravity, but I will use Ceres, Jupiter's moon Callisto, Saturn's moon Titan, and Neptune's moon Triton as examples not because these frozen worlds are suitable but because they illustrate barriers to interplanetary exploration.

The final chapter picks up in the Kuiper belt and finishes with leaving the Solar System. Kuiper objects are distant, dark, and cryogenically cold. These objects sweep out enormous arcs at long periodicities, compared to a human lifetime. Eris, for instance, twice as far from the Sun as Pluto, orbits the primary every 557 years. Our current level of science is dauntingly thin, but perhaps bodies rich in H_2O or O_2 ices could be found and heated to release and capture gases for life support.

The problem of resources is readily illustrated by an island analogy. Before Westerners arrived, New Zealand had no indigenous terrestrial mammals, only two species of bats, and hundreds of species of birds. It had been geographically isolated for so long that terrestrial mammals, having a limited ability to disperse by water, could not reach it. As a result, all of the ecological niches were filled by avifauna that could fly to the islands.

In our case, we are clever terrestrial mammals, but it will be centuries before we will know if we can achieve a high-dispersal status in space. This ultimately depends on the reasons to go and if there are islands with resources. One is also reminded of the old maxim that *islands are places that species go to die*, and we will come back to this idea later.

The distances in the outer Solar System should be daunting enough, but if they are not, consider finding a suitable exoplanet within reasonable proximity—say ten parsecs of the Sun. The continual discovery of new exoplanets also increases the chances of finding an Earth-like planet, but we stand no chance of reaching any of them without traveling at a significant fraction of light speed (c).

One day, my old friend "Q" told me that somebody in the Ukraine had already beamed a message to Gliese 581. It was scheduled to arrive in 2029. Suppose we are actually answered in 2050 and launch a spaceship in 2100 traveling at 0.1 c, which would arrive in the Gliese 581 system in 2300. By 2320, someone here might learn that they had arrived safely. Such travel would require not only rapid transportation but also harvesting and recycling technology (which does not yet exist) that must compensate for *what is missing* in interstellar space. Moreover, intense cosmic radiation is present and punctuated by lethal bursts, and this is nothing but trouble, for now.

Past human migrations, those across the oceans, for instance, have involved not only geographic barriers but also the ability to carry and collect fresh water, gather food, and find islands to raise families. The Polynesians hopped west to east across the Pacific over hundreds of years. Uncharted seas carried the risk of death by storms, dehydration, or stranding without adequate fresh water, fishing reefs, timber, or arable land. Some islands, like Easter Island, lasted only until overpopulation, deforestation, or war did in the inhabitants.

Today, biologists use mathematical models to study the features of animal migrations involving butterflies, salmon, birds, and mammals. The simplest model—*population by diffusion*—predicts that for one species, the number of emigrants falls off as an exponential function of distance from the point of origin. Diffusion has forward and backward rates, but in space back-diffusion would be neglected because turning for home soon becomes impractical.

Diffusion may not predict human dispersal in space because of the large-scale heterogeneities in the locations of islands that could support a colony. Here the distribution of travel distances tends to fall as a power law dominated by the probability of remaining in one place for a long time (Brockmann 2006). Slow dispersal is consistent with spatial inhomogeneity or with periods of consolidation by technological societies, or both. The best predictions use scale-free jumps with long intervals between displacements—perhaps foretelling the future of space exploration, when the reason for the jump outweighs the cost and the risk. However, outside the Solar System, an outpost would be too far away to exchange resources effectively with Earth, and it would be forced to become independent to survive. Even information will be hard to disseminate, and a small colony may not be able to develop its own innovative technologies, causing it to stagnate.

We keep our dreams of space colonization alive by first looking to relativity and then to pie-in-the-sky physics to combat the great interstellar voids. Relativity sets a limit at c, but time slows down, or "dilates," as one goes faster. At 0.999999999999 c, time passes so slowly that in theory one could go to Andromeda, 2.6 million light-years away, and back in one lifetime, though 5.2 million years would have passed on Earth. But the spaceship's mass becomes so great that only the power of a star could push it at even 0.9 c for five million years. So we seem to be stuck in our arm of the Milky Way until someone invents some sort of space-folding trick. Here, let's stop short of dreaming of vacuum-tolerating cyborgs and automatons with artificial intelligence, or of self-replicating von Neumann probes designed to resist radiation, use power sparingly, travel at extremely high speeds, and regenerate themselves, to be sent or decide to go to other galaxies (Macrae 1999). This omission is not because I am uninterested in such mind candy for the daydreamer. I have simply sidestepped these ultrafuturistic speculations in this book because there is enough to accomplish for the time being in our own backyard.

3. THE EXPLORERS

A good student of World War II knows that after the Axis's defeat the United States and the Soviet Union gathered up highly trained German scientists and sequestered them in government laboratories to work on ballistic missile programs. This legendary translocation of scientists, including some avowed Nazis, kindled the breakthroughs in rocketry that produced missiles powerful enough to deliver atomic warheads anywhere in the world with stupefying precision and accuracy. Rocket technology was the heart of the nuclear arms race and the philosophy of nuclear détente that instilled the fear of atomic annihilation into the world. The science of mighty rockets and advanced inertial guidance systems was so highly coveted that the two Cold War rivals worked ruthlessly to steal the secrets of the other while protecting their own.

The arms race, like Janus, had two faces, and the bright side was that these rockets could push past the last vestiges of Earth's atmosphere and escape gravity. The spinoff in rocket technology set the Space Age into motion—all you had to do is put people atop the rocket and launch it into space! The Space Age thus emerged from strange bedfellows: nationalism, which provided the funding; and science, which provided new knowledge.

The story of the Space Race is a fascinating microcosm of the history of science. Being first meant aligning the problem-solving abilities of the world's best aerospace engineers with the sharpest minds in physiology and medicine. Engineers and physiologists would design and fabricate self-contained, lightweight capsules impervious to the near vacuum of space, and a brave explorer with the *right stuff* would climb aboard and ride a controlled explosion into space (Wolfe 2001).

PEOPLE OF ADVENTURE

Explorers push the envelope because it is who they are. The right stuff is often equated with risk-taking behavior, but explorers are usually not self-absorbed adrenaline junkies. They are not conventional risk takers, and it is worth spending a minute to see why.

Explorers set out on their missions for a variety of reasons, including fame and fortune, but they set goals that let them control risk as much as possible. Historically, some explorers have had noble goals, others ignoble, some missionary, and others simply curiosity, but they all share the need to experience a sense of adventure. These reasons are also given by many scientists for doing science. I am no expert on the mind of the explorer, but explorers do not throw caution to the wind; they evaluate risk and uncertainty before and during their quests.

All of us assume certain risks every day, and we will take more risks if the reason is good enough. In other words, our perception of risk is flexible, and the risk-benefit ratio forms the basis for the concept of calculated risk. In spaceflight, the calculated risk is inherently high but varies with the type and phase of the mission, its duration, and on technologies with hard-to-compute failure rates. The worst failures in space, although lethal, have low probabilities and often involve human errors that are hard to anticipate.

Astronauts are accustomed to dealing with inestimable risks, but such times of risk tend to be brief, for example, at launch and during reentry. Periods of uncertainty come in discrete quanta, and the astronaut knows roughly how long each will last. Like most explorers, astronauts develop a feel for the relationship between the likelihood of harm and the duration of risk, but ultimately, the risk assessment is by *gestalt*—an impression of the sum of the identifiable bits of risks.

46

We often also misjudge risk, but most of us adjust our level of comfort with it according to our assessment of the situation and our personal philosophy. We assume more risk if things are bad and if opportunities are limited, for example, as motivation for emigration and colonization. In the future, space explorers may assume open-ended risk with high inherent uncertainty, and anyone headed out of the Solar System faces the risk of permanent disconnection from the rest of humanity.

The idea of uncertainty equates with the unknown, but not necessarily with risk. Faced with the unknown, we are wired for caution—even though exploration, in addition to danger, can produce information that decreases risk. This is another great intangible element of discovery. For instance, the trekker lost in the Sahara may accidentally stumble onto a shady oasis, which may decrease her risk of dying from dehydration or sunstroke before being rescued.

For the sake of argument, I sometimes present my students with an extreme reason for leaving Earth—a backup plan of my old friend "Q." Suppose we determine that life on Earth will be annihilated, say in thirty years, by the impact of a giant comet ejected from the Oort cloud. Assume that 99.9 percent of the species on the planet will be wiped out because dust will block out the Sun for a hundred years and that the Earth will become a dirty snowball. Since humanity would not survive this event, the prediction would provide the impetus to move into space until it was safe to return to Earth. People and their children would climb aboard spaceships, caring nothing for danger or discomfort, in a bid to survive.

Setting that apocalyptic future aside, people will find other, less extreme reasons to leave Earth. The issues in question are the cost, the risk, and the timing. The problem of timing is well known in science fiction, and it is described eloquently by Arthur C. Clarke. Clarke posited how to decide when to send colonists to a habitable planet some light-years away. Take a planet of Gl581, 20.4 light-years away. Should we send a ship that travels at 1 percent of light speed (c) or wait until one is available that travels at 2 percent of c? Since the difference in travel time between the two ships is roughly one thousand years, essentially, *timing* dictates that we should stay put until the faster ship is available, as long as Earth remains habitable. Thus, decisions about space exploration involving people come down to weighing the value of the goal against the costs and risks.

The history of events leading up to human spaceflight gives us perspective on this situation. This history is longer than you might imagine and full of impressive lessons. This is the providence of the pioneer: mistakes sometimes kill irreplaceable people. Those lucky enough not to be in the wrong place at the wrong time should not repeat the mistake. However, apropos Santayana, the lesson lasts only as long as the collective memory. Training and preparation are not domains exclusive to explorers; they figure into centuries of warfare lurking under the guises of logistics, failure-time analysis, survival-time analysis, contingency planning, and emergency preparedness. Preparation is also not about selecting conditions that are just right but about preparing for conditions that are wrong. Experience is the best teacher, but we must also learn from the mistakes of others facing similar challenges.

A famous comment on the experience of the explorer was penned by the illustrious Norwegian polar explorer Roald Amundsen, the first to sail the fabled Northwest Passage in 1903 through 1906 and to reach the South Pole in 1911. Amundsen left medical school at Christiana to become an explorer, and he understood the dangers of polar exploration clearly. Amundsen's thinking on preparation is as apt today as it was a century ago, and his words frame contemporaneous spaceflight: "This is the greatest factor—the way in which the expedition is equipped—the way in which every difficulty is foreseen, and precautions taken for meeting or avoiding it. Victory awaits him who has everything in order—luck, people call it. Defeat is certain for him who has neglected to take the necessary precautions in time; this is called bad luck" (Amundsen 2001).

The last continent, silent Antarctica, had finally been broached before the Great War. The exploits of Amundsen and the British Captain Robert F. Scott became legendary, and like the decades of failed Arctic expeditions, the race for the South Pole was the context for a demonstration of a striking contrast in human survival. Much has been written about why one South Pole explorer succeeded while the other failed, but the circumstances are in general fairly clear.

During the polar summer of 1911–1912, both parties of five covered the nine hundred miles to the South Pole. On December 14, 1911, Amundsen arrived, set his flag, and left a consolation note for Scott's expedition. Scott arrived at the Pole a month later, on January 18, 1912. While Amundsen and his men made it home, Scott and his four companions died on their return trip. Why such a dramatic difference in outcome?

The blame has been laid at the feet of differences in leadership and experience, food and clothing, and the route across the polar plateau. It has been fashionable to blame unusually cold weather as well as Scott's reliance on ill-suited ponies instead of sled dogs. By his own arithmetic, Scott's expedition failed by a razor-thin margin; they were just eleven miles from the final supply depot, on a journey of some 1,800 miles. Still, Amundsen's explanation is best: differences in preparation and experience led Scott to linger too long on the plateau, trapping him and his companions in a great blizzard and starving them to death. So it is that great expeditions rise and fall on their leaders' special insights into contingencies that kill.

POLAR SCIENCE, SPACE SCIENCE

Polar science is one of the grandparents of space exploration, and its history helps us see why human space exploration seems so slow moving. The connection between the two is nicely illustrated by a convergence of events in 1952 that served as a major a catalyst for the space program. That year, the International Council of Scientific Unions (ICSU) proposed an extensive set of global geophysical studies over a period of time that became known as the International Geophysical Year (IGY).

The IGY concept was a reembodiment of the International Polar Years (IPY) of 1882 and 1932, reprising the storied history of international scientific cooperation in polar research. Indeed, the IGY was technically the third IPY and was held on the seventy-fifth anniversary of the first. The IGY brought together the world's leading geophysicists to apply new technologies to gather new information about the Earth. Although space travel was just a dream, rocketry was at the top of a list of important new technologies and brought great excitement to the physical sciences, especially since this was during the depths of the Cold War.

The origins of this excitement are illustrated in the colorful history of the International Polar Years. The first IPY had been the fulfillment of a dream of Karl Weyprecht (1838–1881), a young Austrian Navy officer and Arctic explorer who pleaded the case for international advancement of the earth sciences to the scientific community of his day. He proposed that research be conducted by international teams of investigators stationed at the poles. Weyprecht knew what he was talking about because together with Julius Von Payer (1842–1915), he had led the Austrian-Hungarian North Pole Expedition of 1872 to 1874. For five years after returning home, Weyprecht put forth and refined the idea of a chain of scientific stations encircling the North Pole to gather data on the Arctic climate. At first, his colleagues ignored him, but finally in April 1879, at the second International Congress of Meteorologists in Rome, he garnered the key support.

Karl Weyprecht's single-mindedness derived from his redoubtable experiences on the arduous Austrian-Hungarian Expedition, aboard the 220-ton wooden sailing vessel *Tegetthoff*, named after an admiral under whom young Weyprecht had served. The ship was a barque built in Bremerhaven capable of sail but outfitted with ice-breaking iron hull plates and a hundred-horsepower steam engine. She sailed from Bremerhaven on June 13, 1872, carrying twenty-four men and three years of supplies. *Tegetthoff* sailed north around Norway into the Barents Sea to find the Northeast Passage, an elusive, ice-free passage between the Atlantic and the Pacific.

Weyprecht and Von Payer were proponents of an open polar sea, a popular but untested theory put forth by the renowned German cartographer August Petermann (1822–1878). In the 1850s, Petermann had calculated that an ice-free route could exist in the Arctic sea beyond the pack ice, warmed by the force of the Earth's rotation. Weyprecht and Von Payer wanted to prove his theory by reaching the Bering Strait and navigating through supposedly unfrozen waters northeast of Novaya Zemlya, the icy mountainous islands that separate the Kara Sea from the Barents Sea (figure 3.1). They intended to sail directly over the North Pole.

Petermann guessed wrong, and Weyprecht and Von Payer were beset by ice for two years. *Tegetthoff* slowly drifted north, and in 1873, the ship foundered icebound on the northernmost islands of Eurasia, a mountainous frozen archipelago that lies between 80 and 82°N, less than six hundred miles from the North Pole. The explorers named these islands Franz Josef Land, to honor their benefactor, the Austrian emperor Franz Josef I (1830–1916). But the expedition languished, and Weyprecht and Von Payer finally abandoned ship at 82.5°N. On May 20, 1874, the stranded men left the hemmed-in *Tegetthoff* and spent a long summer returning over the ice to Novaya Zemlya, first in sledges and then in three small open boats. Ultimately, they were rescued by the Russian schooner *Nikolai* and later returned home to Vienna via Norway.

In retrospect, the demands of the expedition had been ill considered, and it fell short of its goals of finding a Northeast Passage and reaching the North Pole. The men were fortunate to survive, but Weyprecht had learned to appreciate not just the value of polar research but how difficult it was. In his mind's eye, he saw a circumpolar ring of research stations, and he finally managed to convince the international meteorological community of their

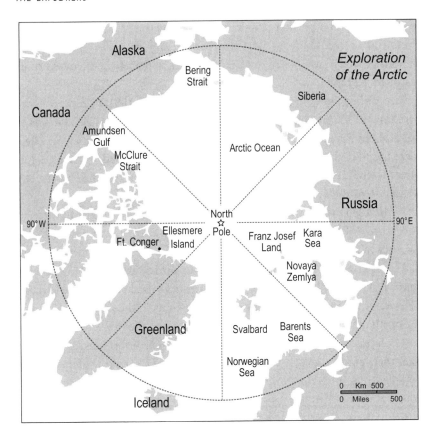

Alaska

Bering
Strait

Siberia

Exploration
of the Arctic

Canada

Amundsen
Gulf

McClure
Strait

Arctic Ocean

Russia

90°W

North
☆
Pole

Ellesmere
Ft. Conger Island

Franz Josef Kara
Land Sea

90°E

Novaya
Zemlya

Greenland

Svalbard

Barents
Sea

Norwegian
Sea

0   Km 500

0   Miles   500

Iceland

FIGURE 3.1. MAP OF THE ARCTIC REGION, SHOWING THE LOCATIONS OF THE EARLY
POLAR EXPLORATIONS THAT LED ULTIMATELY TO THE FIRST INTERNATIONAL POLAR YEAR.

utility. However, a year later, only seven stations were in operation, without help from the United States.

In 1880, at the International Polar Conference in Berne, a plan had been drawn up designating 1882 as the International Polar Year. Everyone was keen to start except the Americans, who seemed reticent. The other nations pressed on, and Karl Weyprecht finally glimpsed success as the circumpolar ring began to materialize. Unfortunately, he never fully realized his dream, for at the age of forty-two he died of tuberculosis, just a year before the scientific work of the IPY began.

The IPY consisted of fifteen expeditions by scientific teams from eleven countries. Twelve went to the Arctic and three to Antarctica to observe the polar auroras, to study geomagnetism, atmosphere electricity, and meteorology.

It was a milestone in international collaboration, but the program was far from trouble free. The poles were treacherous, and life at the remote, frozen stations proved more challenging than had been anticipated even by the most seasoned explorers.

The United States finally got involved, and in July 1881 an American expedition led by Adolphus W. Greely sailed aboard the iron-clad *Proteus* to establish an Arctic base. Their destination was Lady Franklin Bay on Ellesmere Island in the remote Northwest Territories, now called Nunavut. Adolphus Greely, however, did not only intend to make observations. He secretly wanted to be the first to reach the North Pole.

In the 1880s, the Arctic was a place of mystery, steeped in tales of the high adventure of European explorers seeking to transit the frozen top of North America to the Pacific and the Far East. The search for the Northwest Passage had been thwarted by mountainous ice that capriciously crumpled wooden ships, leaving bewildered mariners stranded in a frozen, uncharted world. Blue-water sailors were ill equipped for long stints in the implacable cold of the north and usually met with an untimely end on the ice.

Adolphus Greely and his Lady Franklin Bay expedition fared little better than their predecessors, and the party spent two dark, cold winters plagued by failed resupply efforts, dissension, and ultimately, starvation (Guttridge 2000). None reached the Pole, and inadvertently, the ill-prepared expedition caused immeasurable damage to Karl Weyprecht's dream of international cooperation in polar research.

Ironically, Greely met disaster at Lady Franklin Bay, named for the devoted wife of the illustrious leader of the lost British expedition of thirty-six years earlier, Sir John Franklin (1786–1847). Franklin's party had actually managed to find the Northwest Passage, but none lived to tell the tale. A distraught Lady Franklin, the fate of her husband unknown, funded some forty expeditions to find his two ships, *Terror* and *Erebus*, which had been locked in the sea ice, unbeknownst to her, since 1845.

Franklin's ships were well stocked, and after becoming icebound, the party loaded up their supplies and struck south. However, the men soon became ill, developing insidious lead poisoning from the solder in the tin cans that held their provisions, and by 1847 the entire 128-man company was dead. The combination of unremitting cold, lead poisoning, and malnutrition had clouded their thinking, and poor decisions led them into an escalating cycle of starvation and exposure. This was followed by a tuberculosis epidemic and, ultimately, among the last few survivors, an episode of cannibalism in a final desperate bid to survive.

By 1881, Franklin's expedition was ancient history, and young Greely was a career soldier and a first lieutenant in the Army Signal Corps. He had effusively volunteered to lead the expedition and was appointed by the War Department despite not having a whit of Arctic experience. Greely took twenty-four men, including 2nd Lt. Frederick Kislingbury and the French-trained physician-naturalist Octave Pavy. Pavy agreed, reluctantly, to take a commission as an Army first lieutenant. The government chartered the *Proteus* to convey the expedition from St. Johns, Newfoundland, along the coast of Greenland to Lady Franklin Bay, where a small meteorological station, Fort Conger, was established in Grinnell Land, the northern corner of Ellesmere Island, on August 12, 1881.

Proteus unloaded and decamped to Newfoundland, while the party set about exploring and mapping the world's tenth-largest island, some four-fifths the size of Great Britain. During the expedition, three of the party, D. L. Brainard, J. B. Lockwood, and F. T. Christiansen reached latitude 83° 24' N. This was the highlight and broke the record for "farthest north" set by a succession of British explorers over the previous three hundred years.

Greely had planned to leave the Arctic after a year, but his party became trapped in Lady Franklin Bay by unusually heavy ice. They ran out of food and had to make do with what they could forage. For two consecutive summers, resupply ships failed to reach them, and the men began to die off. In 1883, *Proteus* and her valiant crew did their best to return, but the ship foundered miles short of Greely's camp. Two years after arriving, Greely and his men abandoned Fort Conger in small boats in an effort to rescue themselves.

Leaving in early August 1883, they headed south from Fort Conger, reaching Baird Inlet some six weeks later. They camped at Cape Sabine for nine months and lived on what little food they had carried from Fort Conger—leftovers of previous expeditions and a small cache salvaged from the wreck of the *Proteus*. Greely and his two lieutenants fought bitterly under these harsh conditions. Greely's inexperience in the Arctic had brought serious rebukes from Kislingbury and Pavy, and the furious leader relieved the former of duty; the latter eventually resigned his commission.

By 1884, Greely's expedition had been lost for nearly three years, and finally a relief expedition mounted by the U.S. Navy was sent north in three small ships led by Commander W. S. Schley to rescue the lost men. The Greely Relief Expedition pushed northward through the iceberg seas to Cape Sabine, and on June 22 found Greely and six other men barely alive, one of whom died on the voyage home (Schley 1885). Kislingbury and Pavy died before the ships arrived, and to his family's horror, Kislingbury's body, later exhumed, was found to have been carved up for food after he died. Greely may have eaten Pavy too, yet in the end he emerged a hero. Years later, after a distinguished government career, he was awarded the Medal of Honor for a lifetime of service, by special legislation on March 27, 1935, on the occasion of his ninety-first birthday. He died a few months later.

The North Pole was finally reached in the early twentieth century, but there is stiff controversy about who arrived first. Two expeditions, one by Frederick Cook (1908) and the other by Robert Peary and Matthew Henson (1909) both claimed the distinction, but both have been criticized for lack of proof, and both may actually have missed the mark. In fact, there are some who think Amundsen may have done it first in 1926.

In 1928, long after recollection of Greely's pitiful expedition had faded away and a decade after the Great War, a second IPY was proposed. The world had seen remarkable advances in scientific instrumentation and technology, and in the 1930s, the airplane was already king of the skies. This IPY took place in 1932–1933, on the fiftieth anniversary of the first, and teams from forty nations conducted research in polar meteorology, geomagnetism, and aurora and radio science. The scope of the program, however, was very limited, circumstances that historians have attributed to the economic hard times of the Great Depression. Not surprisingly, the second IPY has been credited with not one single scientific accomplishment.

An interest in geophysical research at the Earth's poles had developed mainly for two reasons: the influence of the poles on global weather was clearly important but poorly understood, and the effects of the spectacular but mysterious polar auroras—borealis in the north and australis in the south—needed scientific explanations. The auroras had a discernable link to solar flares and their interference with worldwide telegraph and radio communications was well known. The northern lights, circumpolar ovals normally extending over much of Scandinavia, Canada, Alaska, and Siberia in the north, bulge south during major solar storms and can easily be seen in the upper continental United States and throughout much of central Europe.

The source of the aurora borealis was unknown until the turn of the twentieth century, when the Norwegian physicist Kristian Birkeland (1867–1917) theorized that the lights were caused by solar emissions colliding with Earth's atmosphere along the lines of its magnetic field. Birkeland had collected magnetic field data from the Arctic and correctly predicted the distribution of electrical charges in the atmosphere over the pole. These observations were ahead of their time but were forgotten for almost half a century until the third IPY, when the physics of the auroras was studied in detail.

The third IPY sprang from a visionary of geophysics, Lloyd V. Berkner (1905–1967), who timed it to coincide with the twenty-fifth instead of the fiftieth anniversary of the second, because he appreciated the import of postwar improvements in technology. The big experiment came to be known as the IGY because it was not limited to polar research. The IGY ran from July 1, 1957, through December 31, 1958, and involved eighty thousand scientists and support personnel from sixty-seven countries at more than eight thousand observing stations. The extended eighteen-month "year" was designed to give the expeditions enough time to establish bases in the northern and southern polar summers and still have an entire year to make observations in both hemispheres (Nicolet 1982).

Lloyd Berkner was born in Milwaukee in 1905 and raised southwest of the Twin Cities, in Sleepy Eye, Minnesota. Much of Lloyd's youth was devoted to becoming what today would be a ham radio or a computer geek. Lloyd studied electrical engineering at the University of Minnesota, graduating in 1927, and two years later found himself flying south on the first Antarctic flight, as radio engineer on the Byrd Antarctic expedition. There he established the first radio link between Antarctica and the rest of the world (Droessler 2000).

Berkner was taken by the science of the ionosphere, and he studied for a Ph.D. at George Washington University but never received his degree. This had little effect on his career, and he repeatedly scrapped for government money for research in the meteorological sciences. Berkner was an unusually winning combination of statesman and scientist whose global view was applied scientifically and politically to the good of geophysics. His powers of persuasion attracted high-level support from scientists and politicians alike in a career that had lasting implications for big science (Needell 1992). Berkner was elected to the National Academy of Sciences in 1948, and in recognition of his lifelong contributions to Antarctic research and to the IGY, an island in the Weddell Sea off the southern shore of the continent was named in his honor in 1960.

The fourth IPY began on March 1, 2007, and ran through March 2009. The program was established by the International Council for Science (ICSU) and the World Meteorological Organization (WMO) and involved more than 225 projects and fifty thousand scientists, students, and technicians in a range of disciplines in the physical, biological, and social sciences from more than sixty countries. In the fifty years since the IGY, geophysics has advanced to the point that modern sensor and computer technology now gives scientists unprecedented temporal and spatial resolution of the Earth (Salmon 2007). Modern polar research utilizes a systems or integrated approach to examine interactions among polar environments, terrestrial organisms, marine life, climate, and other global factors. Most importantly, the fourth IPY was conducted at a time of significant climate change, and the data are an invaluable archive for future polar scientists (Schiermeier 2009).

The most direct route through the Arctic's Northwest Passage, the McClure Strait, was found to be ice free and fully open in August 2007 for the first time in recorded history, thanks to the melting of an area of sea ice roughly the size of California. This long-sought but usually impassable route through the Canadian Arctic Archipelago between Europe and Asia had been discovered by Commander Robert McClure and the men of HMS *Investigator*, who spent 1850 through 1854 searching for the ill-fated Franklin expedition.

McClure had sailed around Cape Horn into the Pacific Ocean and into the Bering Sea, entering the passage from the west. However, like Franklin, his ship was soon icebound and abandoned, and McClure completed the transit overland by sledge. After three years on the ice, he and the other survivors, near starvation, were rescued, and on his return to England, he was court-martialed for losing his ship. Eventually, McClure was vindicated and received ten thousand pounds sterling for proving the existence of the Northwest Passage.

The concentrated research in the Arctic and the Antarctic early in the new millennium has had great import because the poles are changing faster than any other region on Earth, particularly with respect to shrinkage of the ice caps, the accumulation of Arctic pollutants, the holes in the ozone both north and south, and the loss of unique wildlife habitats. Rapid changes in polar climate have global implications ranging from rising sea levels to oil reserves to long-term socioeconomic stability. However important a bellwether the Arctic is, I am rushing ahead of my story.

In 1957, the IGY program in America was the responsibility of a sixteen-member U.S. National Committee (USNC) appointed by the National Academy of Sciences. The USNC operated thirteen technical panels that worked in a dozen areas: aurora and airglow, cosmic rays, geomagnetism, glaciology, gravity, ionosphere physics, navigation, meteorology, oceanography, rocketry, seismology, and solar activity. The thirteenth panel was charged with overseeing the placement of a satellite into orbit around the Earth.

In the United States, several new technologies, especially rocketry, had excited IGY scientists because the new propulsion and guidance systems would make spaceflight possible. As part of America's commitment to the IGY, President Dwight Eisenhower approved a plan to place a satellite into Earth orbit. It would be an American first, and it generated great excitement. Meanwhile, the Soviet Union secretly formulated its own plan to orbit a satellite, and on October 4, 1957, they surprised the rest of the world by launching *Sputnik 1* into orbit. The world's first artificial satellite was the size of a beach ball, weighed just 184 pounds, and carried no instruments. But its effect on the American psyche was profound.

The launch of *Sputnik* created a perception of a technology gap between the two superpowers that stimulated federal spending in the United States on science education and aerospace research, highlighted by the launch of a new agency for federal air and space programs. Congress passed the National Aeronautics and Space Act, and President Eisenhower signed it into law on July 29, 1958. The Space Act formed the National Aeronautics and Space Administration (NASA) from the National Advisory Committee for Aeronautics (NACA) and several ancillary federal agencies. In 1958, Congress also passed the National Defense Education Act (NDEA) to promote education in science, mathematics, and foreign languages by providing low-interest student loans to institutions of higher education. The NDEA also ushered in the modern era of federal support for research in the basic sciences.

On January 31, 1958, the United States launched its first successful satellite, *Explorer 1*, containing a single Geiger-Muller tube built under the direction of the University of Iowa physicist Dr. James A. Van Allen (1914–2006). Van Allen had been an advocate for the IPY from its inception, and his little instrument discovered the broad radiation belts above the Earth that now bear his name. Encircling the Earth in zones containing huge numbers of protons and electrons trapped by the planet's magnetic field, the Van Allen belts are the source of the electrical charges in the atmosphere, and they block much of the cosmic radiation that would otherwise reach the surface of the planet.

The IGY is credited with three major discoveries: the confirmation of the then controversial theory of plate tectonics and continental drift, an accurate estimate of the size of the Antarctic icecap, and proof of the existence of the Van Allen belts. This third discovery has been widely heralded as the first major scientific discovery of the Space Age. As the Van Allen belts are understood today, high-velocity electrons and protons in the solar winds are caught and channeled toward the poles by the Earth's magnetic field. When these charged particles collide with the atmosphere, they excite molecules of oxygen and nitrogen to luminosity, generating the magnificent blaze of the auroras.

NASA had emerged three days before the first anniversary of *Sputnik 1*, with a first budget of roughly 1 percent of the federal budget. The agency's first priority was putting these funds to work to develop a human space program. America's first foray into space was destined to become the most expensive peacetime project in our history.

The launch of a satellite into orbit around the Earth within eighteen years of the end of World War II was no small feat, but it was no surprise either. Indeed, at the secret Nazi rocket test site of Peenemünde on the northeastern Baltic coast, an A4 rocket launched on October 3, 1942, had made a successful ballistic flight to 90 km (56 miles), reaching Mach 5.4 (faster than 4,100 mph). When the Red Army occupied Peenemünde after the war, they discovered plans for three secret rockets called the A9, A10, and A12. The A9/10 combination, euphemistically dubbed the "America rocket," was meant to reach New York. The A12 rocket, like the Saturn V, was extraordinary and, with a proper capsule, probably could have taken astronauts to the Moon.

Shortly after *Sputnik 1*, on November 3, 1957, the Soviets launched *Sputnik 2*, carrying a female terrier named Laika, the first animal in space. The ten-pound Laika was one of ten dogs the Soviets had trained for high-altitude experiments. She could remain in a harness, confined to a small cabin, for a long time. She flew with seven days of supplies, and her vital signs were monitored and transmitted to Earth during part of each orbit. She tolerated the launch but died during the fourth orbit—about six hours into the flight—when her tiny cabin overheated (Malashenkov 2002). Laika's flight, although a space milestone, soon became a rallying point for the humane treatment of experimental animals and did more to promote animal welfare in research than it did for spaceflight.

The National Space Act of 1958 tasked NASA with developing a manned space flight program by whatever technological resources were available, including those of the military. None of the American military services had for-

mal plans for space exploration, but all had missile programs, aerospace technology, and aviation medicine. These areas would come together to achieve
human spaceflight, but first, aerospace engineers and aviation physiologists
had to become partners.

THE HARD-SHELL ENGINEERS

The success of most long forays into extreme environments on the surface of
the Earth is linked to supplies and to a supply line, in some cases with little
margin of error. The deployment of many types of advanced technology is
meant to increase the margin for survival. Every wilderness adventurer is
familiar with snow caves, thermal clothing, freeze-dried foods, water puri-
fication tablets, the GPS, and satellite phones. The technologies range from
the primitive to the advanced and support different aspects of survival in
all but the most extreme terrestrial environments. There is different type of
technology, however, that achieves its ultimate expressions in aerospace and
low Earth orbit, where beyond certain very narrow limits our lives depend
completely on technology. In other words, the only way to visit these extreme
environments is within a *hard shell.*

The hard shell is essentially a noncompressible, gas-tight cabin. The ar-
chetype of the hard shell is the submarine, which operates under conditions
similar in many respects to those of deep space. The stygian depths are cold,
and although the external pressure is high, not low, we must be fully protected by the shell. As in space, resources in the vast undersea are scarce and,
in the deep sea, limited primarily to seawater. In fact, NASA has long seen the
submariner as an analogue of the spacefarer, and it is not a stretch to see the
modern atomic submarine as a prototype starship.

The harsh deep-sea environment greatly limited the diving depth and
undersea duration of the advanced diesel-electric submarines of World War
II, and the need for periodic surfacing to recharge the batteries contributed
to the staggering wartime losses for both the Allied and the Axis submarine
fleets. Finally in the 1950s, the U.S. Navy made a technological breakthrough,
a compact but powerful pressurized-water atomic reactor, under the watch-
ful eye of a brilliant but unorthodox graduate of the Naval Academy, Admiral Hyman G. Rickover (1900–1986). Modern nuclear submarines easily
circumnavigate the globe undersea because the reactor provides essentially

unlimited power for heating, cooling, lighting, oxygen, and potable water. These boats contained the earliest integrated environmental control and life-support systems (ECLSS). Although the basic spacecraft ECLSS design is similar to that of a submarine, the submarine's advantage is not limited to atomic power; it also enjoys a second critical resource: an essentially limitless supply of high-pressure seawater outside the hull.

Submarines desalinate seawater by distillation to provide clean potable water for drinking, food preparation, hygiene, and laundry. Moreover, the electrolysis of seawater provides an inexhaustible supply of oxygen. The left-over brine along with the wastewater is discharged into the sea, where it is recycled by the ocean. Although overboard dumping is also used by spacecraft, the loss of water, because it cannot be easily replaced, has consequences that submarines do not have to deal with.

Submarines also jettison solid waste that sinks to the sea floor, while most garbage in space is incinerated in the atmosphere. Small amounts of debris may sometimes remain in orbit, where it can collide with spacecraft. Importantly, the dumping of material in space also jettisons valuable elements, some of which are reusable and not easy to replace.

Cabin design for submarines is simpler than for spacecraft because it is difficult in microgravity to separate liquids from gases and to collect waste. And in microgravity, pressurized water can be provided only with the use of pumps or bellows driven by high-pressure gases (Wieland 1994). Even natural convection is lacking in microgravity, and the atmospheres in spacecraft must be circulated by fans. Also, the pressure outside a submarine is higher than it is inside, while the pressure inside a spacecraft is higher than it is outside, so gases are not naturally sucked out of the "bubble" of the submarine like they are from a spacecraft. Submarines need little or no N_2 gas to make up for losses to the outside, but spacecraft do.

There is one limitation that submarines and spacecraft do have in common: the means to produce food for the crew. The length of time a submarine can spend underwater is determined by the amount of food that can be loaded onto the boat at the dock. The logistics of submarine patrols depend on storing sufficient supplies onboard to meet the objectives of the mission. This fact impressed me as a young submarine medical officer in New London, where I could look out onto the Thames River and see the black silhouette of *Nautilus* (SSN 571), the first atomic submarine, tied up at the dock. *Nautilus* was just shy of her decommissioning trip to the West Coast after twenty-five years of service. In August 1958, she had become the first submarine to pass under the North Pole, famously broadcasting the message, "*Nautilus* 90 North."

Nautilus reached the North Pole within fifty years of Cook and/or Peary and relied on an advanced inertial navigation system to make the trip entirely underwater. The only danger to her under the Arctic ice, barring an irretrievable reactor scram, or shutdown, was if she had to winter-over for some reason and her crew ran out of food before the following summer. The boat's essentially limitless power for generating oxygen and potable water made her the most advanced warship in history. The expedition was no stunt; it reflected a contemporary philosophy of national defense, the priorities of polar research in the twentieth century, and a demonstration of the power of disruptive new technology.

THE SPACE DOCTORS

A second crucial aspect of spaceflight was an advancement of our understanding of human physiology. During World War I, the U.S. Army recognized this problem and began to train medical officers as flight surgeons. This marked the beginning of formal aviation physiology and medicine and stemmed primarily from the need to diminish the casualty rate of pilots-in-training. The development of a formal curriculum in aviation medicine was the brainchild of Lt. Col. Theodore Charles Lyster, the first chief surgeon of the U.S. Army Signal Corps' Aviation Section, who is today recognized as the father of American aviation medicine. Lyster believed that training doctors in aviation medicine would improve aviation safety. Those physicians became the first flight surgeons, and Lyster's contribution to military aviation not only saved lives but became the basis for modern aerospace medicine.

In 1918, Lyster founded the Air Service Medical Research Laboratory at Hazelhurst Field in Mineola, Long Island, and placed it under the command of Col. William H. Wilmer (1863–1936). Six laboratories instituted research programs on general and cardiovascular physiology, otology, neuropsychiatry, psychology, and ophthalmology. Wilmer's interests were in depth perception and muscle fatigue, and in 1920 he published *Aviation Medicine in the A. E. F.*, a book prepared in the Office of the Director of Air Service. He later worked at Georgetown and in 1925 moved to Baltimore and became the first director of the department of ophthalmology at the Johns Hopkins Hospital. He became the preeminent ophthalmologist of his day and for nine years led the Wilmer Eye Institute at Hopkins, which bears his name.

In order to identify medical factors that would predict the success of new pilots, Army flight surgeons at Hazelhurst Field conducted the first controlled tests of pilot candidates at simulated high altitudes, using hypobaric chambers. To assure the health and safety of pilots operating under the unique stresses of high altitude, these physicians worked hand in hand with aircraft engineers in order to develop procedures and equipment for air safety. Thus began scientific medicine's alliance with aircraft technology, and the field of aviation medicine began its taxi down the runway.

Not surprisingly, the Treaty of Versailles that ended the Great War also nearly ended medical aviation research. In 1919, the Air Service Medical Research Laboratory moved to Mitchel Field, New York, and in 1922 became the School of Aviation Medicine (SAM), signifying the reorientation of the field to the education of physicians in the practice of flight medicine. In 1926, SAM moved to Brooks Field near San Antonio to support operations at the Flying School and to perform medical research for the Army Air Corps. It moved to Randolph Field in 1931, where its tiny staff performed research and promoted education in aviation medicine. In 1959, the school returned to Brooks as a part of the Aerospace Medical Center and in 1961 was renamed the School of Aerospace Medicine to reflect the transition to space medicine.

By the fall of 1927, the Army Air Corps had opened a new facility at old McCook Field in Dayton and named it after the Wright brothers. Wright Field soon became a showplace for military aviation research and, as the airplane evolved from a fabric-covered biplane to an aluminum monoplane, it became the test site for countless innovations in aircraft technology. After Pearl Harbor, Wright Field became vital to America's war effort. Allied aircraft and flight equipment were run through the Dayton facility to measure and improve their performances. Research in aviation medicine conducted at the Aeromedical Laboratory was invigorated by the emergence of problems of in-flight physiological stress from technological advances that allowed the warplane to accelerate faster and to climb higher.

During World War II, some two hundred civilian physiologists and physicians were recruited by the Army Air Corps and the Navy to work on practical problems in aviation medicine. The School of Aviation Medicine at Randolph Field was expanded to accommodate a broad research program in aviation medicine, and the Navy expanded its School of Aviation Medicine at Pensacola and its research base at the Naval Medical Research Institute. These investments paid off in the development and production of oxygen- and pressure-breathing apparatus for use at altitudes above forty thousand feet. Insulated and electrically heated flying suits were also developed for use in confined cockpits and in B-17 gun turrets.

These projects combined advances in engineering with those of anthropometry—the study of body size and shape—in order to design safer aircraft cockpits and better flight gear. These studies also ultimately led to the development of the pressure suit for use at altitudes above fifty thousand feet. Related studies also led to the development of antigravity suits for use in new high-performance fighter aircraft. These innovations, together with basic training in aviation physiology for all American flyers, contributed to the Allied victory, especially in Europe.

The breakthrough for human spaceflight was the development of the pressurized aircraft cabin. The concept had been pioneered in 1930 by the Swiss mathematician and balloonist Auguste Piccard (1884–1962), but it played a limited role in World War II. The penetration of the cabin by flak could cause the type of rapid decompression that had once nearly killed Piccard in his balloon (Piccard 1997).

The first operational pressurized aircraft was the Boeing B-29 Superfortress, which was placed into service in the Pacific at the end of the war. Many high-altitude experiments, including tests of human tolerance to hypoxia, explosive decompression, extreme cold, and high-speed bailouts, were performed in Lockheed Constellation and Boeing C-97 aircraft over Wright Field before the bomber was deployed. Two of the planes, *Enola Gay* and *Bockscar*, delivered the atomic bombs to Hiroshima and Nagasaki.

The results of these efforts were later used by the Air Force to perfect new high-altitude fighters. Many of those involved were career military officers, like Army-Air Corps Lt. Col. Harry G. Armstrong, the founder of the Aeromedical Laboratory at Wright Field and later the first surgeon general of the Air Force. A large number of top physiologists, however, had been recruited to Dayton from American universities, including Loren D. Carlson (1915–1972), David B. Dill (1891–1986), F. G. Hall (1896–1967), A. Pharo Gagge (1908–1993), and W. Randolph "Randy" Lovelace II (1907–1965). They performed the necessary ground and in-flight tests of hypoxia, explosive decompression, and aircraft escape in order to understand the limits of human exposure to altitude (Gagge 1986, Dill 1979).

Lovelace himself performed experiments in high-altitude escape and received the Distinguished Flying Cross in 1943 after he bailed out at 40,200 feet. The parachute's thirty-three-*g* opening shock knocked him unconscious and blew off his gloves, and he suffered severe frostbite to his hands. One of the other great altitude physiologists at Wright Field, F. G. Hall, returned to Duke after the war, and our environmental laboratory here is named after him.

Following the war, Harry Armstrong set his sights on the German aviation scientists who had done landmark research on the medical aspects of high-speed, high-altitude flight for the Luftwaffe. Armstrong collected as many of these physiologists as possible at a center in Heidelberg. In his words:

We took over a building. All the German medical scientists that were willing to work with us were sent to this center; they were given a nominal wage and food, which was in great demand in those days in Germany and each one asked to prepare a chapter or section of a book giving in detail the results of his war time research in aviation medicine. The building . . . was at Heidelberg University . . . later we began sending these people back here . . . altogether we sent thirty-four.

These German aviation physiologists were moved under a secret program known as Project Paperclip, which I learned about in 2002 from two friends, Jay Dean and Rich Henderson, while visiting them in the department of physiology at Wright State. J. D. is a neurophysiologist and a keen historian of aerospace physiology, and Henderson is a Duke medical graduate and career Air Force flight surgeon. Their department chair at the time was the former Duke professor Peter K. Lauf, who I knew years earlier when I was a fellow in respiration physiology. It was old home week for me.

J. D. had been busily unearthing information on the history of aerospace physiology for a book he was writing, and we were discussing high-altitude flight performance during the air war in Europe, when unpressurized aircraft were being flown to ever greater altitudes. As the son of a B-24 armorer flying out of North Africa who had survived being shot down over Italy in July 1943 by a Messerschmitt Bf 109, this was a fascinating topic of conversation for me.

J. D. had a translation of a document describing the origins and structure of the German Air Force called the "Medical and Health Services of the Luftwaffe." It seems that at the time of its secret buildup in 1934 and 1935, the

Luftwaffe had no medical service of its own but soon assembled a formidable cadre of aviation scientists. Among their many efforts, they built mobile altitude chambers mounted on trucks to conduct simulated altitude tests up to forty thousand feet. These chambers were also intended to maintain altitude acclimatization in pilots at the level of a ten-day stay in the high mountains, and the aviators were exposed daily to 16,500 feet of altitude for an hour. These exposures, apart from boosting the pilot's confidence, were probably too brief to be effective, and Luftwaffe crews relied on oxygen masks at high altitude.

The difference in research emphasis between American and German aviation physiologists is fascinating because ultimately the pressurized cabin and the pressure suit, not hypoxia tolerance, would pave the way to spaceflight. The Germans had worked with cabin pressurization in the 1930s, but the Luftwaffe never pursued it. Crude pressure suits had also been used in civilian aviation, but they were too cumbersome for highly maneuverable warplanes.

Ironically, the American physiologists returned to academic life soon after the war (Gagge 1986), while Luftwaffe physiologists brought in under Paperclip went to an expanded aeromedical laboratory at Wright Field to boost the capabilities of the newly independent Air Force and develop the space program. Six German aviation medicine pioneers, including Hubertus Strughold (1898–1986), formerly the director of aeromedical research for the Luftwaffe, were assigned as research specialists to the Air Force School of Aviation Medicine. Strughold was critical in convincing ex-Nazi scientists to come to the United States under Paperclip. Harry Armstrong was made commandant of the school, and by 1948 he and his former enemies were busily contemplating the physiological implications of spaceflight.

In 1920s Germany, well before aerospace medicine was a recognized discipline, Strughold had begun to conduct physiological studies on what he termed the "vertical frontier." During the war, he had conducted studies in acceleration, noise, vibration, and nutrition and later studied jet lag, motion sickness, and problems of space travel, including weightlessness. He invented the cabin simulator and coined the term "space medicine." In 1949, when Armstrong set up the first department of space medicine, Strughold was selected to lead it. He became the first professor of space medicine, eventually earning the moniker of "Father of Space Medicine" (Campbell 2007). In 1963, an award in his honor was established by the American Aerospace Medical Society to recognize excellence in aerospace medicine.

Unfortunately, Strughold had a skeleton in his closet. After the war, he had been linked by the Nuremberg War Crimes Tribunal to medical experiments involving inmates at the first Nazi concentration camp—Dachau, outside of Munich. As the head of the Luftwaffe Institute for Aviation Medicine, Strughold had participated in discussions of human experiments at the institute in the 1940s that included the murder of Dachau inmates by cold-water immersion, exposure to subfreezing cold, hypoxia in low-pressure chambers, and forced ingestion of seawater. Strughold vehemently denied approving such experiments and claimed he had learned of them only after the war. Despite his postwar contributions to aerospace medicine, Strughold's reputation was ruined, and history adjudicated harshly. In 1995, over the objections of several colleagues, his name was removed from the Aeromedical Library at Brooks Air Force Base. In 2006, his name was also stricken from the honor roll of the International Space Hall of Fame (Vorenburg 2006).

Strughold's colleagues in Germany and many of the people with whom he worked in the United States described him as politically disinterested, and he told the American press that the Waffen-SS had threatened his life during the war. This much was true, yet Strughold's hand in the Nazi research network is clearly seen through his connection with the geneticist Hans Nachtsheim (1890–1979), who worked at the Kaiser Wilhelm Institute for Medical Research in Heidelberg (now the Max Planck Institute) on the genetic basis of epilepsy in children. It was Nachtsheim's theory that seizures in susceptible individuals were brought about by cerebral hypoxia, a lack of oxygen in the brain.

In a controversial test in 1943, Nachtsheim exposed six children with epilepsy to dangerous levels of hypoxia in Strughold's altitude chamber in Berlin (Baader 2005). This research on childhood epilepsy was interleaved with human research on convulsions at high altitude, which Luftwaffe scientists believed was also caused by cerebral hypoxia. Indeed, hypoxia does cause seizures at altitude, but the notion of a shared mechanism with epilepsy is incorrect. At the time, Strughold was responsible for this area of investigation for the Luftwaffe, and he would have been required to approve the experiments.

In 1949, space medicine was science fiction to most Americans, but between 1948 and 1951, Armstrong, Strughold, and their colleagues firmly established the conceptual basis for the discipline. Strughold predicted that the majority of the biomedical problems of spaceflight could be defined and addressed within ten to fifteen years and that the necessary hard shells be engineered within fifteen to twenty years. He thought the first manned spaceflights would be possible between 1964 and 1969, a prediction of remarkable perspicuity.

By 1950, the U.S. Air Force had released *German Aviation Medicine—World War II*, a two-volume set prepared by fifty-six leading specialists in German aviation (USAF School of Aerospace Medicine 1950). These books provided detailed coverage of the physiology of high altitude and the problems of acceleration and microgravity. Special symposia were held in 1950 and 1951 by the School of Aviation Medicine in conjunction with the Lovelace Medical Foundation in Albuquerque (now the Lovelace Respiratory Research Institute) to explore topics on the space environment, spaceflight mechanics, and the medical problems of sending people into space. These meetings culminated in a monograph, the *Physics and Medicine of the Upper Atmosphere*, which uniquely integrated astrophysics, aeronautical engineering, aviation medicine, and radiobiology at an applied level. The contributors had recognized the need for cross-fertilization of the key disciplines and were beginning to break down the barriers to putting people into space. Their integrated approach to the new technological and medical challenges became the backbone of space medicine and led quickly to the modern era of spaceflight.

4. TWENTIETH-CENTURY SPACE

The technological history of human spaceflight is heavily chronicled (Compton 1989, NASA 1997, Dick 2007), but this chapter will explain how our knowledge of the problems of putting people into space has evolved and how the remaining issues can be resolved through perseverance, continuity, and international cooperation. Let's begin with a short synopsis of the development of the key twentieth-century concepts that built on the requirement for hard-shell engineering discussed in the last chapter.

THE EARLY DAYS

NASA's seminal manned spaceflight program, Project Mercury, had one objective: to prove that people could survive spaceflight. The obvious first principle was to avoid people suffocating in the near-vacuum of space. To think about the jetliner again: as the aircraft gains altitude, the air outside becomes thinner and thinner as the air pressure falls steadily. In technical terms, the atmospheric or barometric pressure declines geometrically as the altitude in-

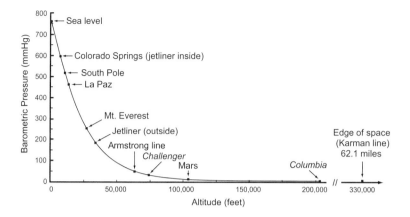

FIGURE 4.1. THE EARTH'S ATMOSPHERE DISPLAYED AS A GRAPH OF ALTITUDE IN FEET VERSUS BAROMETRIC PRESSURE IN MMHG.
Sea level is approximately 760 mmHg (1,013 mbar or 14.7 psi). The atmospheric pressure on Mars is plotted for comparison. At the Armstrong line, water boils at body temperature.

creases above the Earth's surface. This fall in the pressure of the atmosphere from sea level up to the edge of space occurs because the mass of the column of air remaining above you decreases as you ascend. This is why the gas in a helium balloon expands and the rate of the rise of the balloon slows as the balloon floats upward. This fall in atmospheric pressure with a rise in altitude is shown in figure 4.1.

Figure 4.1 contains several key reference points, such as the altitude of Mt. Everest and the pressure of the Martian atmosphere. A pivotal spot is the Armstrong line, named after Col. Harry Armstrong, which marks the absolute limit beyond which the human body cannot survive without a pressure suit. At about 62,800 feet, water boils at body temperature (37°C), producing *ebullism*. The body fluids boil under the heat of metabolism, killing the unprotected flyer within a minute or two. This is why U2 pilots and astronauts must wear pressure suits. The Armstrong line, however, is far below the aeronautical edge of space, the sixty-two-mile Karman line, where the atmosphere is so thin that wings are no longer useful and ballistic flight takes over.

By spring 1959, the United States had announced the selection of the first U.S. astronauts, the "Mercury Seven," who immediately began their training. But almost exactly two years later, on April 12, 1961, the USSR launched the

Vostok 1 from the Kazakhstan cosmodrome, carrying the twenty-seven-year-old Lieutenant Yuri A. Gagarin on a 108-minute Earth orbit, the first man to do so. Within the month, on May 5, 1961, Alan B. Shepard Jr. rode a Mercury capsule from Cape Canaveral, Florida, on a fifteen-minute suborbital flight to become America's first astronaut. Ten months later, on February 20, 1962, John H. Glenn Jr. became the first American to orbit the Earth.

In six flights, Project Mercury provided proof-of-principle for putting astronauts into orbit and returning them safely to Earth. NASA's spaceflight program moved on to Project Gemini, involving a larger new capsule that could hold two astronauts. In ten flights, Gemini provided vital scientific information on weightlessness and allowed engineers to improve reentry and recovery techniques and to establish rendezvous and docking procedures for orbital flights. On June 3, 1965, on *Gemini 4*, Edward H. White Jr. became the first American to leave a capsule in a spacesuit and walk in space. White's twenty-minute excursion was beautiful—but the Soviet cosmonaut Alexei Leonov on *Voskhad 2* had been first, spacewalking over two months earlier. Leonov, however, had nearly died: his suit ballooned up with so much gas that for a time he was unable to climb back into his capsule.

In both the Soviet Union and the United States, Yuri Gagarin's flight bolstered the idea that the Soviets held an edge in science and technology. Gagarin's success was a challenge to America, and Alan Shepard's first flight whetted America's appetite for a space race against its Cold War rival. The battle by the two superpowers in space was fueled by President John F. Kennedy's speech to Congress just three weeks after Shepard's flight. On May 25, 1961, before a joint session of Congress, he announced the goal of "landing a man on the Moon and returning him safely to Earth" before the end of the decade. Kennedy threw down the gauntlet after a series of Soviet firsts: first with *Sputnik 1*, first with a Moon probe, the first man in space, and the first spacewalk. Public backing for Kennedy was strong, and his moon program was overwhelmingly approved by Congress despite that it might cost $40 billion in 1960s dollars.

Over the next decade, the cost of Project Apollo turned out to be about $25 billion, surpassing the Panama Canal twenty-fold and replacing it as the U.S. government's most expensive nonmilitary project. The program was not only more expensive but also more technologically advanced than any program in U.S. history (Siddiqi 2000). Apollo became a poster child for American scientific and technological prowess, and its propaganda value was squarely aimed at the Soviets.

PROJECT APOLLO

Project Gemini ended after *Gemini 12* in 1966, and NASA moved quickly to Project Apollo and a three-man crew. The Apollo system also required three vehicles, including the world's most powerful rocket—the Saturn V. A dual vehicle sat atop the Saturn V, a command-service module combination (CSM) that accommodated a crew of three and a lunar module (LM) with space for two. Once the CSM entered orbit around the Moon, one astronaut remained in the CSM while the others took the LM to the lunar surface, where they set up experiments, explored, and collected rock samples. Afterward, the LM base served as a launch platform for the module's upper compartment to return to the CSM and then to Earth.

The Apollo program was a combination of brilliant successes and horrific failures—bringing respectively acclaim and disparagement to NASA. Two years before the Moon landing, on January 27, 1967, a training-capsule fire at Kennedy Space Center killed the astronauts Roger B. Chaffee, Virgil "Gus" Grissom, and Edward H. White Jr. The men died of asphyxiation from toxic fumes produced by the combustion of synthetic materials in their capsule. The fire had been started by an electrical short-circuit inside the capsule, which was filled with pure O_2 at 16.4 pounds per square inch (psi). At sea level, normal atmospheric pressure is 14.7 psi, and air is 20.9% O_2 (3.07 psi); thus, the capsule contained five times as much O_2 as air.

O_2-fueled fires are ferocious because of the extraordinary rate of combustion. Contrary to popular lore, O_2 is not flammable. Rather, it supports combustion, and if ample flammable material is present, as in *Apollo 1*, an O_2 fire burns explosively. *Apollo 1* burned up so fast that it was impossible for the men to escape from the small capsule. The hatch took at least ninety seconds to unlatch, but the men were dead in under a minute.

The fire prompted a safety redesign of the capsule's electrical system and the decision to thereafter use an air atmosphere aboard all manned U.S. spacecraft. The fire delayed the space program for twenty-one months; *Apollo 7* was finally launched in October 1968. The interlude was used to complete the fire investigation report, implement the changes to the electrical systems, and finish testing the Saturn V rocket.

Meanwhile, the Soviet Union was engaged in a secret but all-out attempt to beat America to the Moon. The West was unaware that some six years earlier the Soviets had suffered an accident similar to *Apollo 1* that had killed a young cosmonaut. In 1961, the twenty-four-year-old Valentin Bondarenko had died in training in a fire in an O_2-enriched capsule, but the Soviet space program was so shrouded in secrecy that Bondarenko's death was not made public for almost twenty-five years.

After the electrical redesign, the Apollo program moved rapidly forward with a flight every few months, and by Christmas of 1968, *Apollo 8* had circumnavigated the Moon. In May 1969, Thomas P. Stafford, John W. Young, and Eugene A. Cernan on *Apollo 10* performed a complete rehearsal of the lunar landing without actually landing. Finally, on July 20, 1969, *Apollo 11*, carrying Neil A. Armstrong and Edwin E. "Buzz" Aldrin, landed on the Moon's Sea of Tranquility, with thirty seconds of fuel to spare. They spent twenty-two hours on the surface, while Michael Collins waited alone in the orbiting CSM for them to return. On July 24, after eight days in space, *Apollo 11* splashed down safely in the Pacific Ocean southwest of the Hawaiian Islands.

Despite that triumph, another accident occurred in 1970, this time on *Apollo 13*, but fortunately no one was killed. On the way to the Moon, some 200,000 miles from Earth, an overheated O_2 tank blew up, and the mission nearly ended in disaster. The tank exploded because in the process of boiling off excess O_2 prelaunch, the temperature became too high and the heater switch fused in the "on" position. The explosion badly damaged the vehicle, the mission was scrubbed, and the crew and ground personnel hurriedly improvised to bring the craft home safely.

Before the launch, several NASA engineers and the mission commander, James A. Lovell, knew that something had gone wrong with the O_2 tank. Agency officials decided not to replace it, and Lovell let it go. He wrote "it was an accumulation of human errors and technical anomalies that doomed *Apollo 13*" (Lovell 1975). Lovell recognized that the accident was a classic double failure: a noncatastrophic malfunction before the mission—a fused switch—led to a catastrophic failure—the explosion of the tank on the mission.

The problem was correctly diagnosed by the *Apollo 13* Review Board, but they soft-pedaled their recommendations. They simply said that NASA "should reassess all Apollo spacecraft subsystems to insure adequate understanding and control of the engineering and manufacturing details of these subsystems," and "where necessary, organizational elements should be strengthened and in-depth reviews conducted"(NASA 1970). *Apollo 14* was held up for nine months while NASA tried to institute these recommendations, but they used a short-term fix, and, to quote Yogi Berra, there was "déjà vu all over again" in the *Columbia* shuttle disaster thirty-three years later.

By the end of the Apollo program, however, the successes had far over-shadowed the failures. In six missions, Kennedy's challenge had been met, and a dozen astronauts had walked on the Moon. In the summer of 1971, the *Apollo 15* crew left a plaque in memory of the fourteen American and Russian astronauts who had died bringing us into the Space Age. Each mission had landed on a different part of the Moon, leaving behind the descent stages of six lunar modules, three rovers, and a variety of instruments dedicated mostly to lunar geology.

When in 1961 the United States announced its plan to land men on the Moon, the USSR had diverted its N1 rocket project to a lunar program in an effort to beat the Americans again. The N1 was the brainchild of the Ukrainian rocketeer Sergei Korolyov (1906–1966), who designed it and in 1956 saw it put a ninety-five-ton payload into orbit using a first-stage cluster of thirty rocket engines fueled by liquid O_2 and kerosene. Korolyov, the so-called secret chief engineer, led the Sputnik program and spearheaded the Soviet space program until 1966, when he died at the age of fifty-nine from a heart attack during an operation for colon cancer. Korolyov's untimely death dealt the Soviets' N1 a death blow, and it never flew. All four test flights from the Baikonur cosmodrome between 1969 and 1972 failed before the separation of the first stage, and the program was canceled in 1974. The design involved too many rocket engines, and at the time the electronic technology did not exist to synchronize and regulate the thrust of so many engines.

A gifted aeronautical engineer, Korolyov had nonetheless suffered horrifically during Stalin's Great Terror. Stalin had him arrested in 1938 for anti-Soviet activities, and he was sentenced to an indefinite term of hard labor. During World War II, Korolyov's teeth were knocked out and his jaw broken during an interrogation, and after two years in the Gulag, he was almost dead. Finally, the Soviet aircraft designer Andrei Tupolev managed to get Korolyov transferred to his design team, where he worked under secret detention. He was not set free until 1945.

After World War II, Korolyov traveled to Germany to study the V-2 rocket, designed by Werner von Braun, the man who would become his nemesis in the Space Race. Initially, Korolyov copied the V-2, but he soon began to formulate his own ideas. Had he lived, the Moon race certainly would have been closer, although with the N1's problems, Korolyov still may not have been able to win it for the Soviets.

The Soviet Union and the United States would not actually cooperate on a truly international program until the Apollo-Soyuz project in 1975. The project worked out rendezvous and docking procedures for the spacecraft of the two countries, and NASA designed and built a docking module that served as an airlock and corridor between the two vehicles. U.S.-Soviet cooperation in space was a big step, but then the United States took a five-year hiatus in manned spaceflight, until the Space Transportation System (STS)—the shuttle—flew its first mission in 1981. Now, with the STS retired, we face a similar lull mainly because of lack of agreement on a plan for a launch system that could overcome the shuttle's limitations. The shuttle's three-element system—the orbiter, an expendable external liquid fuel tank for the orbiter engines, and two recoverable solid rocket boosters—supplied two reusable components that would supposedly cut costs, but in the end they would be the source of trouble.

THE SHUTTLE DISASTERS

Soon after *Apollo 11*, a Presidential Space Task Group set out three space exploration alternatives. The first was an $8 to $10 billion per year program involving a space station in lunar orbit, a fifty-person Earth-orbiting station serviced by a reusable shuttle, and a manned Mars expedition; the second, an intermediate $8 billion annual program, would lead to a Mars mission; and the third, a modest $4 to $6 billion a year program would include an Earth-orbiting space station as well as the Space Shuttle. In 1970, President Nixon opted out because of the cost, and the shuttle, originally envisioned only as the Earth transport component of a far-reaching exploration program, became NASA's centerpiece.

STS-1 launched on April 12, 1981, and demonstrated that the shuttle could take off vertically and glide to a landing without power, and the launch required a rocket with a capacity of only a fifth of the mighty Saturn V. The early shuttle highlights included STS-6 in 1983, when F. S. Musgrave and D. H. Peterson conducted the first shuttle EVA to test a new spacesuit design. On STS-7, also in 1983, Sally K. Ride became the first American woman in space, some twenty years after the Soviet cosmonaut Valentina Tereshkova did it. Finally, in December 1998, the assembly of the International Space Station (ISS) was begun by the crew of STS-88, the *Endeavor* orbiter.

Despite the shuttle's versatility, the iniquitous catastrophes of 1986 and 2003 that killed fourteen astronauts gave the system an aura of unreliability. Both accidents had major repercussions for the space program, and the independent failures of a solid rocket booster and an orbiter, the two reusable components of the STS, called into question the wisdom of the reusable shuttle theory. Indeed, after the second accident, the shuttle did not return to space for over two years, and NASA decided to phase out the aging technology as soon as possible.

The deaths of two shuttle crews drove home the cruelest lesson of human space travel. Attempting to escape or enter the gravity well of an entire planet is a very dangerous thing to do. On January 28, 1986, a leak in the joints of one of the two solid rocket boosters of the *Challenger* (STS-51) caused the external fuel tank to explode in a massive fireball seventy-three seconds after launch, killing all seven astronauts. On February 1, 2003, *Columbia* (STS-107) broke up during reentry over Tyler, Texas, thirty-nine miles up and traveling at 12,500 mph. The left wing sheared off, and the orbiter disintegrated, killing all seven astronauts.

The *Challenger* explosion was investigated in 1986 by the Rogers Commission, which included the iconoclastic physicist Richard Feynman, who deduced that the accident had been caused by a joint failure between the two lower segments of the right solid rocket's motor (Feynman 1988). This failure was caused by a blowout of an O-ring seal that prevents hot gases from leaking across the joint during the burn, and the resulting blowtorch bored a hole through the external tank holding the fuel for the orbiter's engines and ignited it. The commission found no evidence that any other element of the system contributed to the explosion.

The Rogers Commission also concluded that the decision to launch *Challenger* was faulty. At launch time, the temperature was only 36° F, about 15° colder than for any previous mission. A written report by the rocket's manufacturer, Thiokol, had advised against launching at a temperature below 53° F. The decision to launch was made by NASA officials who had no knowledge of previous problems with the rocket booster O-ring joints, and on the day of the disaster, the Thiokol engineers had said no to the launch (Rogers Commission 1986–1987).

The cause of death of the *Challenger* astronauts could not be determined precisely because their orbiter had essentially disintegrated. The fuel tank exploded at 48,000 feet, and the crew compartment separated from the nose cone, cargo bay, and other orbiter compartments. The force of the breakup was not enough to kill the crew, and their compartment continued upward, reaching an altitude of 65,000 feet approximately twenty-five seconds after the explosion. After the breakup, the astronauts' O_2 supply became disconnected from their pressure suits, which would have left them conscious for perhaps fifteen seconds. The crew compartment remained at or above 45,000 feet for almost a minute. As the orbiter spiraled upward, it passed through the Armstrong line, and so began the astronauts' bizarre and horrific death by ebullism: the massive expansion of gas in the tissues that can only be prevented by a working pressure suit.

The crew compartment descended into the Atlantic two minutes and forty-five seconds after the breakup. The velocity at impact was more than 200 miles an hour and far exceeded the structural limits of the compartment as well as the survival limit for *g* forces. Mercifully, the duration of unconsciousness after the breakup was longer than it took for the orbiter to spiral into the ocean. I remember my former Navy commanding officer, the legendary Chuck "Black Bart" Bartholomew, who led the salvage effort for the *Challenger*, telling me later that "the thing hit the sea so hard it broke into a million pieces."

After the *Challenger* accident, NASA grounded the shuttle program for two and a half years while the booster seals were redesigned and new safety procedures developed. The shuttle returned to space on September 29, 1988, with STS-26, and flew eighty-seven missions before the *Columbia* disaster in 2003. Immediately after the second accident, NASA grounded the remaining shuttles, leaving Russia's *Soyuz* and unmanned *Progress* capsules as the only transportation for the ISS.

The *Columbia* Accident Investigation Board (CAIB) reported that two problems had contributed to the crash. The first was technical; the heat-protective tile layer beneath the shuttle had been damaged by the impact of insulation foam that had broken loose from the external fuel tank during takeoff sixteen days earlier. The foam was seen on a shuttle camera during the launch, and NASA engineers had made a note of it. This uncovered a second problem, one belonging solely to NASA—poor communication.

During the sixteen-day mission, despite being informed about the dangers of losing heat-shield tile integrity during reentry by their engineers, NASA managers did not factor the foam into their thinking. On January 23, a flurry of e-mails discussing worst-case scenarios for the loss of heat-shield integrity flew among the engineers at Langley Research Center and the Johnson Space Center. The messages focused on a breach of the left wheel well and the implications of a failure of the landing gear or tires. However, there was no mention of a possible breakup of the craft during reentry.

The CAIB found NASA's response untenable and reproached the "culture" that had allowed safety standards to dwindle at the agency. The problem was made to measure by longstanding poor communication that prevented effective resolution of disagreements. The board, however, also pointed out that even if the agency had assessed the situation correctly, it could not have dealt with a problem that prevented an orbiter from reentering the atmosphere (CAIB 2003).

To appreciate the technical situation, we must see why the orbiter's tile layer was so critical during reentry. The vehicle's nose, the chin, and the leading edge of the wings became extremely hot during reentry, and the maximum heating would occur about twenty minutes before touchdown, when the temperature could reach 3,000° F. When *Columbia* was designed, the orbiter's aluminum and graphite-epoxy skin was insulated with three materials: low- and high-temperature tiles of reinforced carbon-carbon (RC-C) and felt blankets to protect against a wide range of temperatures including a low of −250° F. The felt was later replaced with flexible fibrous blankets and composite insulation. According to NASA, *Columbia*'s thermal protection system contained roughly 24,300 tiles and 2,300 flexible insulation blankets.

The final CAIB report stated that the orbiter was lost because of a "breach in the Thermal Protection System on the leading edge of the left wing" (CAIB 2003). The breach was caused by insulating foam that broke off the external fuel tank and hit the lower half of RC-C panel number 8. On reentry, the breach "allowed superheated air to penetrate through the leading edge insulation and progressively melt the aluminum structure of the left wing." This torch caused sufficient structural failing that the aerodynamic stress of reentry caused the wing to shear off.

The *Columbia* accident, like that of *Challenger*, led to a long delay and the cancelation of shuttle missions. The CAIB report of August 26, 2003, made twenty-nine recommendations for the program and to NASA in general. The board believed that fifteen of these could be implemented before the shuttle returned to flight. In the aftermath, none of the three remaining shuttles flew for two and a half years, and the retirement schedule was set for the program. An obsolescence plan was also agreed on for the ISS, and plans were laid for new space transportation technology.

In January 2004, nine months after the accident, the new Moon-Mars initiative allotted an extra $1 billion to NASA but ordered it to develop a new $12 billion space transportation system over five years by reallocating

$11 billion from existing programs. NASA was told to honor its commitment to the ISS, retire the shuttle by 2010, and have a new vehicle for a manned Moon mission ready between 2015 and 2020. It was also instructed to build a lunar base as a springboard for a Mars mission (NRC 2004). The initial estimate for the cost of the new lunar program, including Project Constellation, was $104 billion.

NASA did respond to the CAIB report and to the Moon-Mars initiative by reorganizing itself (NASA 2004). Predictably, NASA pundits immediately questioned the reorganization, even though its effects would not be known for years. For several reasons, the move was largely cosmetic, but as a result the public din over a "poor solution for cultural problems in federal agencies" finally died down.

The details of the reorganization are less important than the fact that our space program operates on a bloated Cold War government model. It is not clear that the *Columbia* tragedy could have had a positive effect here. The real question is whether the *Columbia* mission was worth the risk and the cost. If anything good came out of the situation, it would be the preference to lay out the goals, costs, and risks of each mission for independent review before approving it. NASA has long been denigrated for downplaying risk, for fear of losing public support, and it is odd that its administrators rarely notice that the agency is hurt by serving a thin veneer of public relations pabulum to America.

On April 14, 2003, NASA did ask an independent advisory group to the NASA administrator, the Stafford Task Group, to assess the agency's plan to put the shuttle back into service. Another task group (the Return to Flight Task Group), was asked to evaluate the implementation of the CAIB recommendations, and it finished up on June 27, 2005, just a month before the flight of STS-114. This was a fourteen-day *Discovery* mission to the ISS intended to check out the new safety measures for future missions.

The Return to Flight Task Group agreed that NASA had met twelve of the fifteen recommendations it was asked to implement before the shuttle flew again. The three most stringent recommendations were not met, but the task group said the agency's extensive testing and equipment modifications showed "substantive progress toward making the vehicle safer" (NASA 2005). The report in hand, NASA reactivated the program.

After the *Columbia* accident, the problem of foam shedding from the external tank continued. The task group acknowledged in their final report, released shortly after the first "return-to-flight" mission on July 26, 2005, that pieces of foam large enough to damage the orbiter critically had again been released. The task group had pointed out that NASA had no in-flight repair capability for the thermal tile but waffled and said the system was "not unsafe."

A disturbing part of the investigation was the harsh criticism of seven members of the task group who dissented in four areas: rigor, risk, requirements, and leadership. They felt that problems in these areas had weakened NASA's ability to manage a high-risk program effectively. They implied that there was only a cursory adherence to best practices and that the agency's management philosophy did not improve after the accident investigation.

The public's perception of the state of affairs at NASA was not helped when the onboard cameras on the STS-114 launch recorded images of foam, one piece three feet long, breaking loose from the external tank. Fortunately, nothing hit the *Discovery*, and the mission ended without a mishap. The foam episode, however, delayed the second "safety" flight of the shuttle for another year, until July 4 through 17, 2006, but after the delay STS-121 went off without a hitch. The end of the shuttle program was plagued by frequent delays, including dents in an external tank caused by a Florida hailstorm. Nevertheless, the shuttle did carry the powerful Alpha Magnetic Spectrometer to the ISS, which has been in operation since 2011. This instrument, together with the Hubble Telescope and its repair missions, are models for how cooperative space science should be conducted. Neither instrument requires an on-site human presence for daily operations, but both can be accessed by people if something unexpectedly goes wrong.

LEO AND THE SPACE STATION

The shuttle program taught lessons through its failures, too: manned space exploration hinges first and foremost on coming and going safely from Earth, which we must learn to do more efficiently. Then, we must learn to travel to the Moon safely, rapidly, and cost effectively. In this respect, aerospace technology, for the foreseeable future, will continue to drive the issue of *getting there*. However, for *staying there*, an established outpost is needed to evaluate the risks of living in space and the practicality of a permanent human presence. Project Constellation, properly managed, would have provided for the implementation of a lunar infrastructure to address the lingering biomedical issues, but it was flawed and put together hastily.

When the Apollo program ended, no one had a blueprint for human exploration, and like today, the way forward was clear neither to NASA nor to the larger scientific community. Apollo was disparaged for its low scientific value, stemming from a failure to engage the broader scientific community and an apparent lack of intellectual payoff. The post hoc debate was grim, but finally, it was agreed to place a large "manned" space station in LEO, where information on "living in space" could be collected.

The rationale for LEO, its proximity to Earth, is unassailable. It is safer and cheaper to supply astronauts in LEO with cargo and to protect them from radiation than anywhere else in space. I tell my students that this is the first great camping trip in the backyard. Years before the first shuttle flight in 1981, NASA had planned a modular station that could be carried up by the shuttle and assembled in space. In LEO, under the umbrella of the Van Allen belts, the station's occupants are protected from cosmic radiation.

LEO is the first rung of the ladder. To push the camping analogy, you don't live in a tent in your backyard for long; you either go inside or, if you like camping, graduate to the mountains as soon as you can. LEO has two other disadvantages: the astronauts are confined to small stations, and they are exposed to microgravity for months.

The history of the space station is fascinating (NASA 1997) and worth a short interlude. The description of an artificial satellite is often credited to Edward Everett Hale (1822–1909), better known for his 1863 short story "The Man Without a Country." Hale published a three-part story in the *Atlantic Monthly* called "The Brick Moon" (Hale 1869), which tells of a 200-foot globe built of 12 million bricks and launched into a 4,000-mile-high orbit by two enormous flywheels quaintly operated by water flowing over a dam!

The purpose of the Brick Moon was to fix longitude for navigators, who could observe it from the sea as it struck out a giant meridian in the sky. In theory, sailors would be able to calculate longitude much like they have for centuries using Polaris for latitude. Hale reasoned that a polar orbit would be necessary but neglected the Earth's continuous rotation beneath his beacon. His "moon" and at least one other would have had to be placed high enough above the equator to remain in geosynchronous orbit (22,236 miles; 35,785 km), allowing mariners to navigate by triangulation.

In the story, some of the workers and their families lived in the structure and were "home" when someone mistakenly activated the flywheels, launching the thing. The hapless people end up in an orbit some 5,000 miles high.

When the satellite is located, it contains thirty-seven survivors who communicate with Earth by jumping up and down on the surface in simulated Morse code. Hale therefore manages to portray the first artificial satellite and the first space station in one fell swoop.

Hale knew nothing of space and only vaguely describes how the inhabitants lived, but his story was prophetic in two ways. He predicted the use of satellites for navigation and that people would use space stations. The longitude problem had been solved in 1761 by the English clockmaker John Harrison (1693–1776) in the form of the precision chronometer (Sobel 1996), but today's geostationary satellites support a highly accurate global positioning system (GPS) for aircraft and ships at sea. A modern handheld or car GPS can place us anywhere on the planet's surface with a precision and accuracy of about a meter.

In its relatively brief existence, the GPS has transformed navigation. It had first evolved for military applications, then for commercial aviation, and finally for everyone else. GPS has become the standard not only for navigation but for accurate knowledge of time and location in space. This is of paramount importance in the geophysical sciences, for instance, in the study of plate tectonics. The success of GPS stems from a concordance of technologies—low-power computing and satellite atomic clocks. Although the U.S. Air Force funded and implemented the GPS, they had objected to developing it for many years. Hard as it is to believe, Air Force administrators cut the program several times, only to have Congress restore it.

The space station was conceived by Russia's famous "Father of Cosmonautics," the mathematician Konstantin Tsiolkovsky (1857–1935). Tsiolkovsky did theoretical work on rocketry even before the airplane was invented, and his novel *Vne Zemli* (Beyond Earth), written between 1896 and 1920, outlined how sustainable habitats could be built in space. Among Tsiolkovsky's ideas were to spin the station on its axis to generate gravity, to use solar power, and to grow fruits and vegetables with hydroponics.

The term "space station" was coined in 1923 by the Transylvanian rocket pioneer Hermann Oberth (1894–1989), who along with Tsiolkovsky and the American scientist Robert Goddard (1882–1945) are considered the fathers of rocketry and astronautics. Goddard's ideas on rocketry were so foreign to the American public that on January 13, 1920, the *New York Times* commented on one of his papers by saying "space travel was impossible, since without atmosphere to push against, a rocket could not move so much as an inch." The *Times* attributed less knowledge to Goddard than that "ladled out daily in high schools." The smug editors, of course, had failed to realize that a rocket pushes against itself.

As a physics student at Heidelberg University, Oberth wrote a dissertation on rocket-powered flight, but it was rejected in 1922 for being too far afield. A year later, it became the basis for his book *Die Rakete zu den Planetenräumen* (The rocket to planetary space). This rejection is perhaps not too surprising, as the 1921 Nobel Prize in physics had been awarded to Albert Einstein "for his discovery of the law of the photoelectric effect." Oberth eventually wrote a second book in 1929 called *Wege Zur Raumschiffahrt* (Ways for space navigation). Both books explained rocket theory, explored its applications, and speculated about space stations and travel to other planets.

Oberth had been inspired by Jules Verne and was a founding member of the German Society for Space Navigation (Verein für Raumschiffahrt). He and his fellow rocketeer Max Valier redesigned Verne's columbiad moon gun to resolve its technical problems, and their Raumschiffahrt gun was intended to fire a 1.2-by-7.2-meter tungsten-steel projectile to the Moon from a barrel some 900 meters long. The barrel would be operated in a near-vacuum to eliminate air compression during acceleration of the shell. To minimize drag, the muzzle would be placed on a 15,000-foot mountaintop, above nearly half of the Earth's atmosphere. This was but one chapter in the checkered history of the "space cannon," which is impractical even for payloads. It would never do for people because the forces of acceleration in the barrel exceed human tolerance by a factor of about a thousand.

Oberth was born too soon, and redesigning Verne's columbiad was a pastime; he had no outlets for his passion until late in life. He considered the effects of spaceflight and zero g environments on the human body and predicted the economic value of satellites. He also envisioned space stations as operating as departure points for the Moon and for other planets, and he proposed a modular design easily launched by rocket and assembled in microgravity.

The first actual technical drawings of a space station appear to be those of a young Austrian army engineer, Hermann Potocnik (1892–1929), who wrote under the name of Hermann Noordung. His early treatise on space-flight, published in 1929 and entitled *Das Problem der Befahrung des Welt-raums—der Raketenmotor* (The problem of reaching outer space: the rocket motor), described his proposal for a space station. Noordung wanted to place a small rotating doughnut (thirty meters in diameter) that he referred to as a *Wohnrad* (living wheel) into a stationary geosynchronous orbit.

In the United States, this idea was championed by Oberth's student Werner Von Braun (1912–1977), the ex-Nazi leader of the World War II German rocket program whose work later catapulted the U.S. space program to the forefront. In the 1950s, Von Braun designed a model of a rotating, spoked wheel to serve as a station with artificial gravity, but it was never enacted for practical reasons. Von Braun's wheel had to be huge in order to minimize the difference in centripetal acceleration between the head and feet of a person standing up in the wheel. The concept was too advanced for the time, but I do remember as a lad how fascinating it was when Von Braun presented it on Walt Disney's *Man in Space* program.

By 1959, a committee of the fledgling NASA had recommended that a space station be placed in orbit in preparation for a trip to the Moon. During the early days of the Kennedy administration, this recommendation was shelved, and by 1969, *Apollo 11* had landed astronauts on the Moon without a space station stopover. That same year, NASA proposed a hundred-man space base, which would serve as a platform to launch nuclear-powered space tugs to the Moon in support of a permanent lunar station. This plan was never realized because of its cost and the mounting paranoia over nuclear power.

The first actual space station—*Salyut 1*—was launched by the Soviet Union on April 19, 1971, two years after *Apollo 11*. The station provided a propaganda boost to a flagging Soviet program and restored some of the prestige lost by its failure to beat America to the Moon. Four days after it was launched, a three-man crew on the *Soyuz 10* spacecraft was sent up, but technical details prevented them from docking with *Salyut 1*, and the crew never entered the station.

Six weeks later, *Soyuz 11* was launched from the Baikonur cosmodrome with the cosmonauts Vladislav Volkov, Georgi Dobrovolski, and Viktor Patsayev aboard. The next day, *Soyuz 11* docked with *Salyut 1*, and the cosmonauts remained at the station for twenty-three days. On June 30, 1971, the men returned to their spacecraft, undocked, and began their return. During reentry, however, communications with them were lost, and when the capsule landed, the three men were found dead, still strapped into their couches.

When *Soyuz 11* separated from *Salyut 1*, a small air valve, which should not have been activated until the capsule was within two miles of Earth's surface, accidentally opened 104 miles (168 km) up, causing the cabin to depressurize suddenly. The miniatmosphere rapidly whistled into space, and the cosmonauts died before they could secure the valve.

Soyuz 11 did not carry pressure suits or an emergency O_2 supply, and once the valve opened, the men would have been conscious for about thirty seconds. As the cabin altitude climbed precipitously, death was by asphyxiation and ebullism. In the aftermath of this accident, the Soviets did not launch again for two years, during which time the spacecraft was outfitted with proper life support. Thereafter, Soyuz carried only two cosmonauts who wore pressure suits for launch and reentry.

The United States' first station was launched in 1973, when NASA used the Saturn V to put the ninety-one-ton Skylab into orbit 270 miles above Earth. Like the Salyut, Skylab 1 was designed to establish and stabilize the station environment before the arrival of the crew. During launch, excessive vibration caused one of Skylab's two solar collectors to be lost and the other to be damaged. Once in orbit, the station was positioned for the remaining solar panel to generate as much electricity as possible, but the inside temperature climbed to an intolerable 126° F (52° C). The problem was resolved by rotating the station to decrease its solar profile and having the crew of Skylab 2 deploy umbrellas. Skylab ultimately supported three missions in which three astronauts lived onboard for twenty-eight, fifty-nine, and eighty-four days, respectively. The station was abandoned in 1974 and fell to Earth on July 11, 1979, scattering debris across the southern Indian Ocean and western Australia.

It hadn't taken NASA long to realize that orbital stations were prohibitively expensive, and by 1982 it had established a Space Station Task Force that eventually proposed an international partnership to share the cost and responsibility for a new station. In 1984, NASA was authorized to build an orbital station, but the rationale, design, and cost of the proposal stirred up a firestorm, and plans for a more modest facility called Space Station Freedom were released in 1991. The project was again scaled down in 1993 by the Clinton administration, and the resulting three-quarters sized orbital facility became known as Space Station Alpha.

At the time, the Soviets were focused on long-duration space missions, and they eventually launched a series of seven Salyut stations. On February 20, 1986, they launched the first element of a new station called Mir (Peace), ultimately composed of seven modules. A decade later on April 26, 1996, the seventh and final module was placed into orbit. Soyuz was used to transport crews and cargo to and from *Mir*, and during its fifteen-year lifespan the station made more than 86,000 orbits and was visited by 104 astronauts from several countries, including the United States.

Mir did shine, but it was also fraught with notorious mishaps. A *Mir* highlight occurred on January 8, 1994, when the *Soyuz* TM-18 spacecraft carrying three crewmembers arrived, including the physician Valery Polyakov, who remained onboard *Mir* until March 1995, the record for a single mission. In theory, the 437 days he spent in space was long enough for Polyakov to make it to Mars.

After the collapse of the Soviet Union, the Russian Space Agency could no longer afford to maintain *Mir* and attempted to commercialize it. By early 2000, it was clear that the venture was going nowhere, and the systems and hardware on board were beginning to degrade. Metal fatigue in the older modules of the station had become a serious problem. In mid-2000, *Mir*'s last crew departed, and the decision was made to deorbit the station. In 2001, *Mir* plunged through the Earth's atmosphere and disintegrated over the South Pacific Ocean somewhere near 40°S/160°W.

Russia's long experience with prolonged spaceflight on the *Salyut* and *Mir* stations was offered to the new American-led effort, and eventually some Russian hardware was included. In 1993, this led to the decision to build the new station jointly and call it the International Space Station (ISS). In 1994, Russia and the United States undertook the first of nine shuttle-*Mir* missions, and seven American astronauts lived aboard *Mir* in order to provide collaborative experience for the construction of the ISS.

The ISS became a unique partnership of sixteen nations, including the United States, Russia, Canada, Japan, and the nations of the European Space Agency (ESA). More than forty space flights were planned over five years, using at least three different space vehicles—the shuttle and the Russian *Soyuz* and *Proton* craft—in order to place station components into orbit. The ISS construction was greatly delayed by the *Challenger* accident, but NASA completed the bulk of the station before retiring the shuttle. Originally, the ISS was expected to last ten years, and the first Russian segment, or RS, was designed to last for fifteen years. A crew of six, staying for six months, would have been the norm, with short-term support for up to twelve. This capacity was largely determined by the ability to remove CO_2 from the station's atmosphere.

The ISS design was based on a contemporaneous life-support philosophy, including the level of systems redundancy. The Russian and U.S. approaches to redundancy and safety have always been different, particularly with respect to the backup systems for critical life support functions, such as O_2 production. If the main O_2 generator fails, for example, the U.S. protocol calls for a second identical system, whereas the Russians use bottled O_2 until the generator is repaired.

Such differences reflect a different philosophy of conserving mass and reutilizing resources. According to a 1998 technical report from the Marshall Spaceflight Center, NASA's life-support philosophy was derived from the cost of placing large payloads into orbit (Wieland 1998). The principle is to minimize the use of expendable materials and to utilize technologies for regeneration, for example, in removing byproducts like CO_2. Another goal is to recover as much mass as possible—to minimize the loss of mass, for example, by recovering water from the air during the removal of CO_2 and returning it to the cabin. A third aspect is to save weight by reducing the built-in redundancy to within reasonable safety limits because redundancy is expensive and backup equipment still requires maintenance.

NASA's approach requires that certain life-support equipment be designed *never to fail*. Accordingly, on the ISS, the tolerances for many life support (ECLSS) functions were set to zero. However, nothing is perfect, and a zero-failure design cannot be applied to primary life-support equipment because no matter how unlikely, a failure is usually fatal. The *Soyuz 11* tragedy, with cosmonauts without pressure suits, is a case in point. For critical functions such as cabin pressure, O_2 concentration, and temperature, redundancy must be maintained, and NASA "criticality ratings" guide the implementation of this rule.

In November 1998, the first two modules of the ISS were launched and joined in orbit. Other modules soon followed, and the first crew arrived in October 2000—the U.S. astronaut Bill Shepherd and the Russian cosmonauts Yuri Gidzenko and Sergei Krikaley, aboard a *Soyuz* spacecraft. Since 2000, the ISS has been inhabited continuously by rotating crews. The experience of setting up, monitoring, and maintaining a station before the working crew arrives is invaluable in planning a future moonbase.

Finished and operational, the current ISS has the length and width of a football field, a mass of 816,000 pounds, a module length of 157 feet (51 meters), and a truss span of 357.5 feet (109 meters). Its habitable volume is 12,705 cubic feet, roughly the same as a five-bedroom house or the cabin of a Boeing 747 jet. There are six laboratories that provide far more research space than any previous station. The ISS orbits at an average altitude of 354 kilometers (220 miles), at 51.6° inclination to the equator.

Before the *Columbia* accident, ISS cargo missions were planned for approximately every three months using U.S. shuttles or Russian *Soyuz* and *Progress* capsules. Nonreusable items were loaded into *Progress* return vehicles, which disintegrate on reentering Earth's atmosphere, or into an orbiter for return to Earth for disposal. Typically, the shuttle ferried long-term crews, while *Soyuz* carried short-term astronauts, cosmonauts, and paying customers to the ISS. But after the *Columbia* accident, Russian spacecraft became the only form of transportation to and from the ISS until 2006.

In 2004, the frequency of resupply visits to the ISS was reduced because of a shortage of Russian cargo rockets, and the crew size, which had been three, was reduced to two. Russian spacecraft are smaller than the shuttle—a new *Soyuz* capsule carries three passengers and a *Progress* capsule about 2.5 tons of supplies. In contrast, the shuttle handled seven people and twenty-five tons of cargo.

The smaller number of cargo missions to the ISS led to significant maintenance problems as well as to the accumulation of junk that would normally have been taken off by spacecraft returning to Earth and jettisoned into the atmosphere to burn up. This problem brings into sharp focus yet another problem a Mars mission will face in supporting five or six people for 900 days without resupply. Certainly no one would want to see humanity's first mission to another planet leave a massive pile of junk on the ancient surface of our pristine neighbor!

These few selected points from the history of spaceflight clearly indicate that *getting there* and *staying there* are different issues. Technologically, both should be getting easier, but they are both becoming too expensive. Before the ISS, NASA's exploration program devoted most of its resources to getting there and back safely instead of staying there. The Russians spent most of their time figuring out how to stay there. Both sets of lessons are important. The rest of this book is focused on staying there, beginning with the Moon, the only place large enough and close enough to implement a long-term strategy for interplanetary space exploration involving people.

5. BACK TO THE MOON

America spent $3 billion 1972 dollars for each of the twelve astronauts that went to the Moon—an extraordinary outlay, given the meager scientific return on the Apollo program. This was not lost on the scientific elite of the 1970s, and thirty years later, the Moon-Mars initiative found itself in the same boat, and cost overruns and poor planning prevented it from weathering its own storm. Americans have expectations of a real return on the space program, but no one really seems to know what this means.

Part of the problem is the conquest mindset, a holdover from the Cold War, which confuses things. Switching focus from the Moon to an asteroid fits that pattern and plays on a the-sky-is-falling mentality. It gives the impression that we are acting in the best interests of the world, but the logic is weak. The Moon fell out of favor because of the canard that we have been there before. This specious argument has not gone unnoticed by many present and former NASA people, including former astronauts and mission directors, who are publicly shaking their heads.

Over the Independence Day weekend of 2011, I watched NASA Administrator Charles Bolden on C-SPAN speaking to the National Press Club about the future of manned space exploration. He was cagey about NASA's new heavy-lift plans, which hadn't quite materialized at that point. Bolden recognized this as the most important decision he was likely to make at NASA, and as I watched him I realized we weren't going back to the Moon—or to any asteroid, either—by 2025. But if it performs as advertised, our new space transportation system should eventually provide plenty of flexibility and muscle.

The Moon-Mars initiative would have monopolized a large share of NASA's resources over twenty years. When astronauts did return to the Moon, they would have visited briefly to collect geophysical data, because we have no technology for a long stay on an airless Moon. This technology gap cannot be closed by sending a few people to an asteroid or to the Moon for a few days. Ultimately, if we are going to stay in the spacefaring business, expensive as it is, we will have to invest in a moonbase.

In 2006, NASA unveiled a lunar exploration strategy that involved setting up a small base at the lunar south pole and having it staffed permanently by 2024. Physical science research appeared to drive the plan, but the ultimate idea was to use the base to develop a springboard for Mars. A range of activities called "lunar exploration categories" were listed, each with several objectives. A synopsis of the plan is shown in table 5.1.

Not surprisingly, the plan emphasized infrastructure and other things that NASA does well, with little emphasis on the people, even though it required them to live there for long periods. The original timetable, about eighteen years, is too short to transport and install semipermanent infrastructure from LEO sites to the Moon. Also, nascent surface technologies needed for Mars must be "incubated" on the Moon.

The selection of the rim of the Shackleton crater near the south lunar pole as the site for a lunar base was driven in part by the long nights at the Moon's equatorial latitudes, which cuts the solar irradiance in half and deepens the thermal cycles. These temperature cycles cause composite materials to expand and contract continuously, causing early fatigue. These effects are lessened at the poles because of the more uniform temperature caused by nearly constant sunlight. The Moon has no area of perpetual summer, since its 1.5° axial tilt is appreciably smaller than Earth's 23°, but the polar region in the south has short nights suitable for a solar-powered base. And 122 kilometers from the pole, the summit of Malapert Mountain, five kilometers high, receives full or partial sunlight for 93 percent of the year and stays in Earth's line of sight, for good communications (Sharpe and Schrunk 2003).

TABLE 5.1. CATEGORIES OF LUNAR EXPLORATION.

FIELD OF EMPHASIS	EXPLORATORY CATEGORY	# OF OBJECTIVES
Physical Sciences	Astronomy and astrophysics	9
	Solar physics	8
	Geology	16
	Earth Observations	12
	Material Science	3
	Environmental characterization	12
Life Sciences	Human health	8
Operational	General Infrastructure	8
	Environmental hazards mitigation	5
	Habitat, life support, monitoring	11
	Operations, quality control	4
	Power	3
	Comunication	4
	Position and navigation	5
	Surface mobility/transportation, crew activity	12
Commercial	Lunar resource utilization	11
	Development of commerce	23
	Global partnership	4
	Historic preservation	4
	Public engagement and inspiration	13

Source: http://www.gov/pdf/163560main_LunarExplorationObjectives.pdf.

The site at Shackleton crater would also provide a *cold trap* at the crater's base (Sanderson 2007). A base initially housing four people for two weeks would eventually fill a role similar to Antarctica's McMurdo Station, which the National Science Foundation uses to stage research across the continent. From the base, traveling astronauts could set up instruments and collect lunar samples.

A small base is the first step in the development of permanent lunar infrastructure, but the McMurdo analogy is a bit of the old NASA soft shoe, as a lunar base would be far more expensive than even the Amundsen-Scott South Pole Station, some 850 nautical miles south of McMurdo. The idea of rotating personnel through the station on a two-week basis is also dead in the water, even using the ISS as a waypoint. The mantra that the Moon is a stepping stone to Mars is stating the obvious; learning on *and about* the Moon before setting a Mars timetable is the real trick. The lack of a lunar scientific directorate was naïve and even made reelection-seeking politicians nervous.

This oversight and bringing Mars into the picture opened the door to the criticism that money would be wasted getting ready for a conquest in lieu of durable, useful research.

NASA did identify some critical research areas, including low-frequency radio astronomy, lunar interactions with Earth's magnetosphere, electromagnetism, radiation, and the effects of dust exposure on people and equipment. Other science was shelved, such as optical astronomy, basic geology, gravitational waves, astrobiology, and cosmic rays. The decisions were based on an estimated return on investment by science administrators, who are notoriously bad at predicting profitable spinoffs.

In the early 2000s, Russia's RKK Energia had proposed an ambitious lunar exploration program that included a permanent moonbase, but the plan never received government support. After NASA's 2006 announcement, the Russians promptly expressed interest in working with the United States, offering expertise in booster rockets and other hardware in exchange for help by Russian scientists in staffing the base. Russia's interest in international cooperation was a common-sense move given the ultimate cost of the enterprise and the continuing effects of the 2008 economic recession.

Russia has also brokered space deals with the European Space Agency and with China. The Chinese purchased durable Russian *Soyuz* technology and have been flaunting their own space program since China launched its first taikonaut in 2003. China's ambitions are the purview of the semiautonomous China National Space Administration (CNSA) and are rolled up into a three-phase program called Shenzhou ("Divine State") that originally began in 1992 as Project 921. Shenzhou culminates with the launch of a space station into LEO, targeted for 2020, which coincidentally dovetails perfectly with the retirement of the ISS. The Chinese version would have a mass of sixty metric tons, less than one-fifth of that of the ISS, but it would be the sole space station in operation at that point.

The lifespan of the ISS can be extended until at least 2030, and NASA has pushed hard for that extension to avoid relinquishing dominion over LEO to China. China commonly heralds its peaceful intentions through space cooperation agreements with more than a dozen countries. These agreements have led to joint projects, mostly with Russia, although the United States is kept at arm's length. This estrangement is the fault of both countries, but it has raised few eyebrows in Washington because China is still seen as decades behind the United States in space technology. The 2012 NASA budget even explicitly forbade cooperation with the Chinese, thanks to the "Wolf clause," inserted by a U.S. Republican congressman from northern Virginia who thinks he can block espionage and decrease human rights abuses in China by not cooperating with them in space.

The activities of the CNSA are closely tied to the Central Military Commission and embraced politically by the Communist Party of China. Although backroom government concerns about the possible militarization of space by China are perhaps unduly suspicious, both China and the United States have amply demonstrated their prowess at bringing down "obsolete" satellites using ground-based missile systems.

More concerning was the 2010 announcement that the CNSA plans to send taikonauts to the Moon by 2025. The CSNA espouses a long-range goal of establishing a permanent manned observatory on the Moon, which Chinese space experts acknowledge as a necessary step toward the exploration of interplanetary destinations including Mars. They have also announced tentative plans for a manned Mars mission around 2050.

Any moonbase will be enormously expensive, and despite China's growing economic power, it is not clear that the nation can fund a base without strong international partnerships. If they are serious, it means that they will be investing in a new space transportation system for the Moon and Mars. By going it alone, China is quietly positioning itself to take the lead in human space exploration should American resolve falter in the face of a soft economy. And their political system can dictate the expenditures necessary to establish a lunar presence, which if successful, would make it easy to lay exclusive claim to lunar mineral rights, over international objections. This is reminiscent of the international elbowing brought on by climate change over the ownership of valuable Arctic resources, like oil, that are becoming harder to extract elsewhere on Earth. The Moon has a long list of surface resources, including aluminum and titanium; hence, lunar mining for the Chinese could become economically viable by the end of the twenty-first century.[1]

SERENE SELENE

The Moon has been written about more than any other celestial body, yet when I ask my students if our Moon has an official name, only a third of them recall "Selene," the beautiful mythological Greek goddess sister of Helios. Despite Selene's dimpled beauty, she is barren and hostile. The average distance between the Earth and the Moon, about 238,855 miles (384,400 km), is roughly ten circumnavigations of Earth at the equator. This is a long way, but the ISS covers this distance in about fourteen hours; it is not the barrier it once was.

The Moon is tidally locked; the same hemisphere faces us throughout its twenty-seven-day orbit. Its size and proximity affects us profoundly, for instance, in the generation of the tides. It has a gravity of 1.62 m/s^2, about one-sixth of Earth's. This "partial gravity" is a decided advantage of over stations in microgravity and makes the Moon a huge, stable orbital platform with invaluable resources for human space exploration (Duke et al. 2003). When it comes time to depart, her escape velocity is only 5,325 mph (2.38 km/ second), compared with Earth's 25,000 mph (11.2 km/second). Lunar escape is also helped by its diaphanous atmosphere—there's virtually no friction. These physical features are as good as it gets for learning to live in space.

A lack of curiosity about the Moon is also a bit hard to understand, considering that it is roughly 2,160 miles (3,475 km) in diameter—and was once part of the Earth. Its natural history, especially its relatively large size compared with Earth, has long intrigued planetary scientists. Most "moons" in our Solar System are far smaller than their planets. A number of theories have tried to explain this, such as the old capture theory (Hartmann 1997; Murdin 2001, s.v. "Giant Impactor Theory"), but the Apollo lunar samples have changed how we think about the Earth-Moon system.

Selene was knocked off of the proto-Earth some 4.5 billion years ago by a cataclysmic collision, while our planet was still molten. This is the "giant impactor" theory, even though no one knows what hit the Earth. It explains the Moon's low density (3.3 grams per cubic milliliter, or 60 percent of Earth's) and lack of an iron core. It also explains the identical O_2-isotope composition in lunar and Earth rocks and the high mass of the Moon relative to Earth (1.23 percent of Earth), compared with the smaller moons of Mars (Georg 2007).

How much of the Moon derives from Earth and how much from the ancient interloper, called Theia, is unknown. Theia would have been like Mars in size, and its fate is unknown (apart from the fact that it collided with the early Earth), but the Moon is less than a tenth the mass of Mars. This may mean that multiple collisions occurred and that most of the smaller body was absorbed into the Earth. Theia probably had a different chemical composition than the proto-Earth, but the Moon's isotope composition is identical to ours. So the impact may have peeled off part of the Earth's crust to form the Moon. The Moon is also lopsided, with most of its mountain ranges on the far side, and no one knows why.

The Moon was originally much closer and thus once had a larger effect on the Earth, but the tidal forces of each on the other are still considerable. Apart from the tides, the Moon has stabilized the Earth's tilt and evened out our seasons. The tidal effect of Earth on the Moon may contribute to the moonquakes recorded by passive seismometers placed on the lunar surface during the Apollo missions. Minor seismic activity from thermal cycles and meteorite impacts would be expected, but those seismometers have recorded deep moonquakes, source unknown, a few exceeding 5 on the Richter scale. This is strong enough to endanger a lunar station (NRCCSCEM 2006).

The Moon not only has no meaningful atmosphere; it has no liquid water. This exaggerates the daily temperature variation on the surface, which is roughly -250 to $+250°F$, about twice the diurnal variation on Earth. The absence of air and water, the temperature swings, and cosmic radiation make the Moon a dangerous and inimical place.

Apart from the threat of radiation, these conditions on the Moon are similar to those outside the ISS, and they can be managed similarly by thermal engineering. The absence of "air" means that heat exchange occurs primarily by objects heating up by absorbing short wavelengths of light and cooling off by emitting longer infrared radiation. On the ISS, a combination of highly reflective insulation and radiators with large surface areas are used to dissipate heat and maintain the internal thermal stability.

The Moon's wispy atmosphere was first measured in 1972 using a mass spectrometer left by the last lunar astronauts, Eugene A. Cernan and Harrison H. Schmitt, on *Apollo 17*, in the Apollo Lunar Surface Experiments Package (ALSEP), which operated for five years. The atmospheric pressure on the Moon at night was found to be about one ten-billionth of the pressure on Earth at sea level (3×10^{-15} bar or 2×10^{-12} mmHg, compared with Earth's 760 mmHg). This is essentially a vacuum and no defense against cosmic radiation.

The evanescent atmosphere is composed mostly of four gases in roughly equal proportions: neon (^{20}Ne, ^{22}Ne), helium (^4He), hydrogen (H_2), and argon (^{40}Ar, ^{36}Ar). There are traces of CO_2, ammonia, and methane (CH_4) but almost no O_2. The entire mass of the lunar atmosphere is a mere 25,000 kilograms (55,000 lbs), less than that of a tanker truck. All of the O_2 in the Moon's atmosphere would support four people on a moonbase for about a week.

In addition, the Moon lacks a magnetic field to shield it from bombardment by galactic and solar rays, such as ultraviolet light (UV), and dangerous sun flares comprising intermittent solar particle events (SPE) and coronal mass ejections (CME). On Earth, we are protected from this radiation by the Van Allen belts and by our dense ozone- and nitrogen-rich atmosphere. The radiation flux on the Moon is hundreds of times greater than it is on Earth, and it waxes and wanes with the eleven-year solar cycle. On the Moon, we would need overhead electrostatic generators or water-jacketed habitats to repel high-velocity charged particles—or live ten feet underground.

When a large SPE strikes the Moon, it packs a wallop that would do serious harm to an astronaut. Large solar storms carry high-energy protons of more than one hundred million electron volts (100 MeV) that can easily pass through six inches (fifteen cm) of water. As early as 1989, the National Council on Radiation Protection estimated that radiation sickness from a major SPE could kill an astronaut caught on the lunar surface in a spacesuit. Even modern spacesuits offer little protection against such intense radiation.

WATER, WATER EVERYWHERE

There is no atmosphere; extreme radiation, heat, and cold; dust; and moonquakes—but for all those things, the apparent lack of liquid water on the lunar surface was for many years considered the most troublesome issue in lunar exploration. The absence of water was often cited as a reason for never going back. In the early days of lunar observation, the many large dark spots on the Moon were thought to be seas and were named accordingly. By 1900, that romantic illusion had evaporated, and by the 1970s, the analysis of the Apollo lunar rocks indicated that they, and the Moon from which they had come, were, for all practical purposes, dry. But it turns out water really is there, in the soil, or regolith, and at the bottom of deep craters, though still unavailable in a conventional sense. The distinction between available and unavailable water may seem like a fine point, but the development of affordable water-recovery technology could govern the establishment of a permanent lunar base.

The situation was so doubtful that some scientists thought bulk water would have to be transported from Earth and those costs factored into the cost of operating a base. Although a new specialized delivery vehicle could reduce the cost dramatically, water would still cost thousands of dollars per liter, essentially the price of gold. In 2006, S. Alan Stern, the lead scientist for the *New Horizons* spacecraft and a former NASA scientific director, proposed a lunar water plan. Stern had been a fiery opponent of Pluto's demotion and was known for creative thinking in the space sciences. He and his team at the Space Science and Engineering Division at the Southwest Research Institute proposed launching ice payloads from the Earth directly to the Moon, a plan euphemistically called the SLAM. Stern's group calculated that only about 15 percent of the ice in a proper SLAM would be vaporized at typical impact speeds and that most of it would stay within five feet (1.5 meters) of the surface. The problem is that one cubic meter of water weighs one metric ton (ice weighs slightly less), and the payload must be landed at night and recovered before lunar sunrise to keep the ice from sublimating into space.

On the other hand, some planetary scientists had thought that eons of meteor and comet impacts had deposited so much water ice over the lunar landscape that it could not have all been vaporized by the impacts or dissipated into space. Ice could be trapped at the bottom of deep craters, in the dark shadows under the rims, where cold and darkness would prevent its sublimation. Such cold traps would be an invaluable source of water, which could be split into O_2 and H_2 for both life support and rocket fuel. However, problems with collecting the appropriate lunar data prevented the question from being answered of whether either pole harbored a practical amount of useable water ice.

The evidence for lunar ice remained inconclusive, despite the use of orbital instruments capable of sophisticated radar and neutron measurements. These instruments detected ice-compatible signals, but whether this was actually ice or simply hydrogen was hard to say. The first such signal, detected in 1994 by the Defense Department–NASA *Clementine* Moon probe, was radar evidence of ice in the permanent shade at the bottom of a crater near the south lunar pole, where it could persist at extremely low temperatures, roughly 100 K. In March 1998, NASA reported that the neutron spectrometer aboard NASA's *Lunar Prospector*, which operated in lunar orbit for eighteen months, had detected hydrogen deposits at both poles.

This polar "ice" was originally thought to be spread over 10,000 to 50,000 square kilometers (3,600 to 18,000 square miles) in the north and 5,000 to 20,000 square kilometers (1,800 to 7,200 square miles) in the south, but later calculations showed that this area was roughly 1,850 square km (650 square miles). The ice mass was crudely estimated at some 6 trillion kg (6.6 billion tons). The level of confidence in the estimate, however, was low because the measurements were crude.

When the *Lunar Prospector* failed to yield to a definitive answer, the probe was crashed into the area of the strongest signal, in an attempt to pick up a water-vapor plume after the impact. The crash failed to throw up water but left the possibility open that the area of the impact contained up to 1.5 percent water, distributed over a large area. The *Lunar Prospector* couldn't make out structures smaller than about thirty miles across. Moreover, the powerful radar at the Arecibo Observatory in Puerto Rico had found no signs of ice at depths of up to one meter. This led some scientists to conclude that water was trapped at only very low concentrations and only in the top few meters of the lunar soil. This would make it difficult and expensive to recover.

In 2003, the European Space Agency (ESA) launched the *SMART-1* satellite as part of a program to test inexpensive new space technologies. Once in Earth orbit, *SMART-1* fired up an ion engine, spiraling away and spending twenty-two months circling the Moon. It probed deep polar craters using an advanced camera to map variations in *albedo*, the fraction of the light falling on an object that is reflected off its surface. The albedo can be used to infer certain things about the chemical composition of the surface. *SMART-1* also had an infrared spectrometer to search for the spectrum of water ice, but instead it found the signatures of calcium, aluminum, and magnesium in the lunar rock, not water ice. *SMART-1* was sensitive and should not have missed water.

In 2008, the floor of the Shackleton crater (four kilometers deep), which is permanently shaded and was predicted to hold ice deposits, was imaged with the ten-meter-resolution terrain camera aboard the *SELENE* spacecraft as it was faintly lit by sunlight from the top of the rim (Haruyama et al. 2008). The estimated floor temperature was less than 90 K, cold enough to trap ice; however, the albedo suggested that pure ice was absent or that an ice-soil mixture covered a small percentage of the area.

Lunar scientists continued to focus on the polar craters because the probability of finding ice there was still higher than anywhere else, and in 2009 they learned of water ice near the Moon's poles from the joint *Lunar Reconnaissance Orbiter (LRO)/Lunar Crater Observation and Sensing Satellite (LCROSS)*. Launched on June 18, 2009, the system was to collect data on possible landing sites and map water ice distribution in the upper layers of regolith in circumpolar craters.

The *LRO* remained in a polar orbit thirty-one miles (fifty km) above the surface, while the *LCROSS* sent the Centaur rocket, followed shortly by an instrument-bearing shepherd spacecraft, crashing into a deep southern polar crater, called Cabeus, excavating tons of material. Analysis of the plumes for water vapor indicated significant ice at the site, up to 5 percent by weight (Colaprete 2010). Equally surprising was that *LCROSS* dug up a lot more than ice; it found CO_2, methane, and ammonia. Meanwhile, the ultraviolet spectrograph on the orbiting *LRO* observed signs of carbon monoxide, mercury, calcium, and magnesium (Gladstone 2010).

Earlier, using a different approach in 2008, India had sent the *Chandrayaan-1* spacecraft to the Moon to map the surface with an imaging spectrometer. It wasn't long before analysis of the maps began to show large areas of water concentrated mainly at the poles (Pieters et al. 2009). This observation was confirmed by a reexamination of older data from the *Cassini* spacecraft's lunar flyby on its way to Saturn and later by the *Deep Impact* spacecraft (Sunshine et al. 2009).

In short, lunar scientists now seem to agree that there are three kinds of water on the Moon. There is water on the surface formed from the solar wind, "deep" water from previous volcanic activity, and comet water left over in deep craters, which are actually colder than previously thought—in some places colder than Pluto. However, the evidence so far is indirect and would be strengthened by recovering core samples to be returned to Earth for analysis. Cores of material would provide a history in ice and reveal the presence of organic compounds delivered to the Moon over billions of years. Geochemical and isotopic analyses could tell us about the origins of lunar ice, the composition of impacting bodies, and where in the Solar System they originated. On the other hand, if the signal is mainly attributable to the solar wind, then we must learn to turn molecular O_2 and hydroxyl (OH) drawn from lunar soil into water.

The initial cost of water recovery technology will be high, but cost-amortization of the technology with that needed for splitting out O_2 and for efficient life support and recycling will help lunar exploration become more feasible. Nonrecyclable and lost resources must be replaced either from Earth or from lunar sources, which is very expensive. Fortunately, the lunar soil is rich in metal oxides from which O_2 can be recovered (figure 5.1). Learning to use indigenous resources on the Moon is a critical step for interplanetary exploration, because resupply missions from Earth will never be practical for Mars or missions to the outer Solar System.

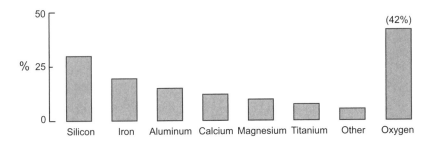

FIGURE 5.1. THE MAJOR METAL OXIDES PRESENT IN THE LUNAR REGOLITH. Overall, O_2 is 42 percent by weight.

IT'S NOT MADE OF GREEN CHEESE

Despite the bad conditions, the Moon is not made of green cheese, and although some aspects of establishing a base there are on soft ground, the soil and ice are primed for *in situ* resource utilization (ISRU, in NASA speak). As NASA folk so fondly say, the Moon is a great test bed for ISRU. ISRU will be discussed in more detail later: it is a hot topic at NASA think tanks, especially for O_2, H_2O, propellants, and construction materials. The Moon is a natural laboratory to develop new technologies for collecting and refining indigenous resources for use in exploration when Earth is too far away to provide them. The Moon's real value, as I tell my students, is like the value of real estate: location, location, location. This value will increase as the logistics of going back and forth gets easier and safer.

NASA is also banking on the Moon's partial gravity and lack of an atmosphere to provide for easier disembarkation for Mars. The Moon is an outpost in our own backyard, but it is also close to sites useful for "storing" things in space for transport to other locations such as asteroids and Mars. The two Lagrange points L1 and L2 are in near-stable equilibrium orbits with the Sun-Earth-Moon system. The positions are metastable for weeks to years, so that supplies can be cached and retrieved later. Earth even has its own little Trojan asteroid at a Lagrange point. And L1 is home to *SOHO*, the *Solar and Heliospheric Observatory Satellite*, where it enjoys an unobstructed view of the Sun, and L2 is to be the home of NASA's new James Webb Space Telescope (Hanover et al., 2006).

The Moon's large flat surfaces and stable gravity make it more convenient and less expensive to recycle water, grow plants, build or dig shelters, and stay clean. These are important quality-of-life issues. The Moon can be used to store, deploy, and service solar panels, electrical grids, and scientific instruments. It has even been suggested that expanses of regolith can be modified into a *glace* to collect sunlight. Moreover, the underground capacity is essentially limitless, accessible to robotic mining, and safe from radiation.

The Moon has up and down directions. Liquids behave normally; water flows, and neither fuel nor a gyroscope is required to maintain the position or orientation of a lunar outpost. And partial gravity will become the de facto norm for all human space exploration in the Solar System. This must be studied carefully because partial gravity has important implications for missions to Mars and beyond.

The benefits to musculoskeletal health of living in partial gravity instead of microgravity should be easy to establish. At one-sixth gravity, bone and muscle loss should be slower and easier to prevent with exercise than it is in microgravity, where efforts to counteract atrophy can be time consuming. Exploration-class missions in microgravity will involve multiyear stints followed by long periods in the partial gravity on Mars or a moon of one of the outer planets. Anywhere in the Solar System we might feasibly visit has less gravity than the Earth. In fact, except for Mars, which is 0.38 g, gravity would be similar to that of the Moon, and some places, like Ceres, have even less gravity. In order to set the horizon just right, we must understand partial gravity as a barrier to human space exploration.

Another problem, a reliable source of power, has multiple potential solutions, but these must be reduced in practice. The Moon and the Earth receive the same average solar energy or irradiance, about 1,360 watts per square meter (1.36 kW/m^2). On Earth, this energy is concentrated at the Equator, and depending on atmospheric conditions, about a quarter of it is scattered and absorbed by our atmosphere. This is not the case on the Moon because of the lack of an atmosphere to absorb solar energy. In contrast, Mars receives less than half the solar energy of Earth, and maximum irradiance at noon on Mars is about 43 percent that of the Earth (595 W/m^2). Under extreme conditions, such as dust storms, this can fall to 7 percent of the equatorial maximum on Earth.

If America's lunar exploration program is revived quickly, we will lead in the development of technologies to recover and process raw materials from the lunar regolith, for instance, for construction. The regolith contains large deposits of ilmenite—iron and titanium oxide ($FeTiO_3$), which is 32 percent

O_2 by mass. Heated with a catalyst, ilmenite will release large amounts of O_2. Powder and stone fragments of regolith can be sintered (fused in a mold) or melted to create different products, for instance, by using microwaves. Focused microwaves may also be useful for shielding, splitting water, and making solar cells and pavement from lunar soil.

Some years ago, I was wandering around at a national meeting on lung diseases and stumbled into a workshop on Moon dust. I came away with a healthy respect for it. The lunar regolith is ubiquitously layered with fine dust, and it has many troublesome features. It consists mainly of irregularly shaped pieces of impact-produced glass; each shard has a large surface area laced with jagged edges (Park et al. 2008, Liu et al. 2008). It is iron rich, highly electrostatic, and sticks like Velcro to everything from spacesuits to electronics.

The particle size borders on ultrafine, ranging from 100 to 300 nanometers. Such fine airborne dust on Earth would have serious effects on lung function, including the distal air sacs, or alveoli. Our lungs would probably not fare well breathing this dust, even in trace amounts, especially over months in partial gravity, which tends to keep it suspended in the atmosphere for longer periods. On Earth, such particulates are considered air pollutants, and they are some of the causes of major lung diseases, including lung cancer (Knaapen et al. 2004). Dusty trades such as mining produce disabling chronic lung diseases, such as silicosis, and there is abundant silica in Moon dust.

The key to minimizing exposure to lunar dust is to keep it out of the habitat. NASA is designing habitat-egress procedures that would leave the spacesuits in an airlock or coupled to it, away from people and equipment. A suit would never actually return to the habitat once it has been used on the surface. Another approach is to "dust" the suits with special antistatic vacuum cleaners. Solutions to the dust problem are critical to Mars exploration too, where long-lasting planetary dust storms are commonplace.

The idea that lunar resources cannot be converted inexpensively to useable commercial products has long been popular with the naysayers, but it is losing traction. An example often used by the former *Apollo 12* astronaut, geologist, and U.S. senator Harrison H. "Jack" Schmidt is the notion of mining helium-3 (^{3}He) on the Moon. Schmidt thinks ^{3}He may be exportable for use in fusion reactors and that mining it could produce useful byproducts from lunar soil for life support and propulsion, including O_2, H_2, and H_2O (Schmidt 2006). He has dreamed up a cooperative solution between the U.S.

government and the private sector that does not directly involve NASA. He envisions a competitive return on investment using low-cost commercial vehicles designed to harvest the regolith, with living quarters for the crew and a large capacity for bringing ^{3}He-enriched material to Earth for refining. To the science fiction buff, the idea is agreeably reminiscent of spice mining in Frank Herbert's classic *Dune* novels.

Schmidt thinks ^{3}He mining could be a financial "carrot" for commercial use of the Moon by opening the door to cheap fusion power and escaping the use of fossil fuels on Earth. ^{3}He is a stable light isotope of ^{4}He embedded in the lunar soil by the solar wind. The amount of ^{3}He in the top three meters (ten feet) of lunar soil is estimated at about fifty kilograms per square kilometer. This could power a thousand-megawatt fusion reactor for a year: a few truckloads of lunar soil could thus power the United States for a year. Schmidt argues that the engineering problems can be solved in the next few decades to make the entire process viable.

Schmidt's proposal to fuse ^{3}He with deuterium and tritium, heavy isotopes of hydrogen, is feasible, but ^{3}He is not the preferred fuel of fusion experts. It is more difficult to fuse with deuterium than is tritium because helium's extra proton raises the fusion temperature fourfold. Conventional fusion devices called Tokomaks create hot, dense plasmas in confinement vessels to fuse deuterium and tritium. However, more energy is required to heat the fuel than is released by the reaction—the energy breakeven point has not been reached. So Schmidt's idea is not soup yet, but the man himself has been to the Moon, and he sees the big picture. We will learn to capitalize on scientific and commercial opportunities on the Moon.

Scientists building the most advanced fusion reactor in the world, ITER (Latin for "the way"), in the south of France want to reach the breakeven point in 2020. The problem with ITER is that the Tokomak is huge; it requires, among other things, 10,000 tons of supercooled magnets and a powerful vacuum. You could envision generating power for a lunar colony by placing a Tokomak in a crater on the Moon, where it is already cryogenically cold, but for spacecraft, nonconventional low-mass confinement vessels, for instance a system that fired atomic nuclei directly into one another in an electrostatic field, would be necessary.

I have no illusions that fusion will work on the Moon, but there are many other useful resources in the regolith. I mentioned the ilmenite deposits, located mostly in the lunar lowlands, as a source of O_2, but the iron and titanium are useful too. Many other metal oxides in lunar soil will yield pure metals, like aluminum, after the O_2 is extracted. The collection of lunar metals will preclude having to transport them from the Earth. Furthermore, our proclivity to deplete rare-earth metals here on Earth may one day make it economically necessary to mine lunar metals and export them home. This raises serious societal and ethical issues over the right to exploit lunar resources, and they represent significant challenges for the future.

Earth is facing shortages of certain metals, including platinum and titanium, for which the discovery of major new reserves seems unlikely. There is also a list of rare-earth elements, such as hafnium, gallium, iridium, neodymium, tellurium, dysprosium, europium, terbium, and yttrium. Most people know nothing about these but they are part of everyday things such as computer screens, touchscreens, solar cells, and high-performance magnets (Crow 2011). Some are crucial for the success of emerging green technologies, but there are few known deposits, and recycling technology is limited. Some are already in short supply in part because China accounts for most of the world's rare-earth mining, and they restrict the supply on the world market. Those elements that are not yet critical may become so within thirty to fifty years, and by 2100, the world could also face shortages of copper, aluminum, tin, and zinc (Cohen 2007).

Lunar aluminum and titanium, coupled with the discovery of surface deposits of platinum, iridium, and other critical elements, could make it more cost effective to export them to the Earth than to extract them from the Earth. In either location, implementing *sustainable* technologies requires greater capitalization than simply finding something disposable, but the need for durable technologies to harvest and recycle such materials for major consumables is inevitable.

The idea of putting high-value lunar resources into the commercial realm sparks strong feelings about protecting the Moon from destructive special interests. In 1967, an Outer Space Treaty, and in 1979, the United Nations Declaration on the Moon (Moon Treaty) were drafted but never ratified by the main players, the United States, Russia, China, and Japan. The objective was to prohibit the placement of military hardware on and ownership of the Moon and other Solar System resources by private citizens or corporations and to encumber off-world exploitation of nonrenewable resources, similar to the International Law of the Sea and the Antarctic Treaty.

The use of lunar materials for habitat construction is generally considered fair game. Safe haven from space radiation is of paramount importance, and the possibilities range from the inflatable torus to lunar subsurface tunnels.

There are many conventional and exotic construction alternatives, but lacking practical experience, NASA has been making "educated guesses" from among a wide range of possibilities. The first lunar habitats will likely involve tough inflatable cylinders or donuts connected by airlocks, similar to large hyperbaric chambers. These would incorporate a central core for electrical and life-support equipment and perhaps small water-jacketed rooms for protection from solar storms. However, I would personally prefer to go underground when the Sun cranks up.

Perhaps the most efficient of several possibilities for permanent living is that of clearing out subsurface lava tubes left over from the Moon's ancient volcanic history. This is also a fabulous geological research opportunity for lunar scientists who are curious about what lies under the regolith. We hit two birds with one stone by using the advanced robotic mining technologies already being deployed here. Remote-controlled mining, known as *teleportation*, is used to protect miners from exposure to heat and other hazardous conditions. Teleportation eliminates travel to and from the mine, which for the Moon is quite a distance. Teleportation is also used in blasting and is being developed for deep seabed mining. It offers the means to revolutionize our most difficult and important mining ventures.

The application of underground tunneling, sealing, and gas pressurization to a lunar outpost was first proposed in the 1970s, and since 2000, similar proposals have been on the drawing boards for Mars. These are critical developments for space exploration because we will always need robots to do the dirty work in the Solar System. Mars probes already have performed miniature excavations, but large-scale tunneling is real work and takes huge amounts of power. Electrically powered tunnel-boring machines, however, are already state of the art.

Robotic mining also takes astronauts in cumbersome spacesuits out of the picture. Once again, the cost and efficiency of robotic work must be taken into account, but viewing old film clips of Apollo astronauts shuffling clumsily about on the Moon illustrates the problem nicely. In the 1970s, lunar spacesuits were "top-heavy"—the center of gravity was too high, and the astronauts tended to fall over easily, which is dangerous. Suit technology has been improved to minimize top-heaviness, but durability, joint mobility, and glove flexibility still need work. More fiddling time is needed to perfect this critical human interface with the Moon, but the harvesting of lunar resources will fall almost exclusively into the realm of the robot.

EXTRAVEHICULAR ACTIVITY

In going back to the Moon, we will need four types of hardware: habitats, rovers, robots, and spacesuits. Prototype habitat technology is available, and there are new ideas aplenty. A new enclosed electric rover built by NASA could go sixty miles and, according to my old friend, the astronaut Michael Gernhardt, is essentially ready for a test ride. Today's spacesuits work fairly well for extravehicular (EVA) activities in microgravity and allowed NASA to climb the infamous "Wall of EVA" in building the ISS, but exploring the rough lunar terrain will require a rugged new spacesuit design.

Spacesuits are stiff because they have a high internal pressure relative to the near-vacuum of space. The ISS (as did the shuttle) operates at a cabin pressure of one atmosphere (14.7 psi, 760 mmHg, or 1,013 mbar). In other words, U.S. astronauts breathe air at a pressure equivalent to that at sea level. However, operating a spacesuit where the internal pressure is 14.7 psi and the external pressure is essentially zero is like trying to move an inflated tire from the inside. The pressure must be lowered (and the O_2 concentration increased) in the suit to reduce the pressure difference across the garment, making it less stiff. The current suit systems for EVA operate at 4.3 psi, or roughly 0.3 of an atmosphere.

In order to get into a suit at lower pressure, the astronaut must "climb" from sea level to the equivalent of 30,000 feet—about a thousand feet higher than Mt. Everest. Raising the O_2 concentration prevents hypoxia, but the astronaut faces the same risk of decompression as a scuba diver during ascent: *decompression sickness*, or the *bends*. The bends occurs when N_2 gas dissolved in the tissues comes out of solution and forms bubbles, which may cause pain, numbness, or weakness. To avoid the bends, astronauts must undergo lengthy periods of O_2 prebreathing with exercise before EVA in order to eliminate excess N_2 from their bodies.

Lunar habitats will operate most effectively at a mountaintop altitude; in other words, at a lower pressure than at sea level. Lowering the habitat pressure makes it possible for an astronaut in an O_2-filled spacesuit to exit the habitat to an even lower working pressure more quickly, because less N_2 must be eliminated beforehand to avoid the bends. This also decreases the volume of gas lost to the outside when airlocks are opened to move in and out of the habitat.

For argument's sake, let's place our lunar habitat at the altitude of Colorado Springs (6,035 feet, or 1,840 m). This would allow an astronaut in an O_2-filled spacesuit at 30,000 feet to make a faster egress because the amount of N_2 eliminated from the body moving between 6,000 feet and 30,000 feet is only two-thirds as much as starting at sea level. If the habitat atmosphere is enriched in O_2, the pressure can be kept even lower, making egress even easier. More O_2 also avoids another problem—altitude or mountain sickness, which can produce disabling headaches, nausea, vomiting, and sleep disturbances (West 1998).

The air in Colorado Springs has an O_2 partial pressure (PO_2) of approximately 128 mmHg, which translates in someone with normal lungs to an arterial O_2 saturation of about 95 percent. In addition, the O_2 concentration in a lunar habitat could be raised safely from 21 to 26 percent, which would provide an inspired PO_2 similar to sea level and decrease the amount of N_2 in the atmosphere by 5 percent.

Habitat pressures as low as 0.55 ATA (8 psi) and O_2 concentrations as high as 32 percent are being considered seriously by NASA, thereby creating an atmosphere the same as that at 5,000 feet of altitude. The use of more O_2 and less N_2 (68 percent instead of 78 percent) would allow barometric pressure to be lowered to the level of 19,000 feet. This greatly decreases the time for decompression to 30,000 feet because compared with sea level only 44 percent as much N_2 must be eliminated. This might actually avoid the bends without any decompression time.

Putting these ideas into practice means integrating them into the all-important habitat, but electrical engineers hate low-pressure environments because electronic components tend to overheat. If you have driven down the mountain from Pike's Peak, you will recall that the ranger stops you every few miles to check the temperature of your brakes. The engineer must design, install, monitor, and maintain low-pressure-habitat electronics once the basic specifications are agreed upon for atmosphere control, shielding, recycling, reliability, and durability. We know low-pressure habitats work; the South Pole Station, for instance, safely supports scientists at an elevation of 9,300 feet (2,790 meters), but the altitude equivalent, because of polar atmospheric lows, can be as high as 10,800 to 13,330 feet (3,240 to 4,000 meters).

All habitats, from submarines to skyscrapers, are mechanical systems that ultimately break down and wear out, especially where moving parts are involved. Equipment on the Moon will not be exempt from this principle, particularly since the hardware will be constantly exposed to radiation, dust, and deep temperature cycles. The reality of this problem was brought into the public eye in 2004 and 2005 by the troublesome failure of the O_2 generator on the ISS, a famous test of the limits of life-support reliability and durability.

On a permanent moonbase, spare parts from Earth will be limited because high cost will restrict cargo missions to only a few times a year; however, Russian *Progress* spacecraft have proven that resupply missions can be handled robotically. Thinking about Mars, cargo missions will be even more expensive, and the constraints of orbital mechanics may drastically restrict even unmanned missions. Given the state of the art, without more efficient propulsion, life-support system durability will be tested for decades to come.

The dependability of space hardware has been aided by the short duration of missions and their proximity to Earth. Since Apollo, every mission has been in LEO and supported from Earth. For the ISS, cargo craft are stocked on the ground and launched to the station. Cargo craft obviously have a smaller size and lower cost than manned ships, but the failure to maintain this lifeline is potentially disastrous, and the ISS was put in jeopardy in 2003 when the *Columbia* disaster grounded the shuttles for more than two years.

On the Moon, the optimal habitat technology will be simple, redundant, robust, self-monitoring, and perhaps self-correcting because first outposts will not have a cadre of professional engineers. And space life-support systems, just like the HVAC system in your house, must be maintained, updated, and replaced as the technology advances.

Beyond the Moon, the resources will be limited mainly to what the astronauts bring with them. Even the cache on a gigantic ship is limited in comparison to a moon. The isolated spacefarer must fend for herself, and only by traversing regions rich in harvestable matter or by reaching some fertile new "island" can this problem mitigated. I will say some more about such new islands later.

The life-limiting resource is O_2. Without it, we remain conscious for mere seconds. If someone sapped the O_2 out of your room instantly, your head would hit the desk before you finished reading this paragraph. For space technology, if a malfunction involves the loss of the capacity to generate O_2, the clock starts ticking, and a backup system must be brought online quickly.

A tight system has essentially no leaks, but there are never zero leaks in space. The O_2 supply and its recycling must be watchfully monitored because if you cannot regenerate it or get home before it runs out, the effect is lethal. The probability that compromise of a critical resource like O_2 will wipe out an ISS crew is lower than the same malfunction killing everyone on a lunar outpost because the ISS is close enough for the crew to escape to Earth or for NASA to rescue them. From this standpoint, today's life-support systems are ready for the Moon but not for Mars. In the future, new technologies for Mars, like integrated semiautonomous systems, can be installed and tested on a lunar outpost.

In addition to integration, the issue of scale is important but not so easy to quantify. Biologists know that only habitats of a certain size create stable ecological systems, but without information, it is not easy to set scale requirements for space habitats. For instance, until actual measurements were made, no one knew that disconnected plots of Amazon rainforest below an area of a few hectares suffered an appalling and rapid loss of species diversity.

Scientists naturally squabble about the minimum size of space environments and only seem to agree that the problem needs attention. The philosophy of miniaturization and intensification of hardware is deeply ingrained, yet downsizing is more than an engineering issue; it is a biological one. Issues of scale are handled from a dual perspective of the characteristics of the systems technology and the human factors. The picture is fuzzy, and finding a scale that is "just right" takes fiddling time. We are lucky to have the Moon on which to do it.

THE VIEW FROM EARTH

The environment at the bottom of our warm, O_2-rich atmosphere is vastly different than anything the Moon, despite its terrestrial origin, has ever seen. We evolved to have a rapid metabolism, using O_2 and hydrocarbons to produce energy and generate heat. Because we need heat to function, heat management is a biological problem. Our body temperature on the absolute or Kelvin scale is 310 K (37°C), and our tolerance range is narrow.

On the Moon, the usual temperature range is 100 K to 400 K, but in the shadows, the temperature falls precipitously. Certain deep craters have temperatures near those on Pluto (40 K). Practically speaking, most biological reactions stop at *the freezing point of water,* where the phase changes to ice (273 K; 0°C). This phase change can be circumvented in some species, like the Antarctic ice-fish, whose blood contains antifreeze or ice-resistant bacteria that freeze-thaw (or desiccate-rehydrate) easily and regrow when favorable conditions are restored. The highly specialized cells and tissues of large organisms, like mammals, do not fare so well with large swings in the temperature of their environments. We can overcome this limitation within reason as long as we can breathe and drink, because our intelligence allows us to adapt behaviorally to heat and cold. Behavioral adaptation is our only option for living under the extreme conditions on the Moon.

The heat distribution in space is uneven, and liquid water is a rare commodity. Next to oxygen, our lives are defined by liquid water, and it in turn is defined by the temperature. We evolved in the Sun's habitable zone; if we move too far from the Sun, we dissipate more heat into the environment than we can collect from it. Life's order arises thermodynamically from the concentration of the star's energy, at the cost of universal disorder or entropy. Heat flows from hot to cold, and once the temperature equilibrates, it can only be raised by adding more heat. Outside our habitable zone, this leaves us just two options: live inside insulated spaceships and habitats using a self-contained source of power until we reach another habitable zone, or transform a cold place to suit us by causing it to accumulate heat.

These options are not mutually exclusive. Both require behavioral adaptations to effect a change in environment, but one is practical, and the other is not. We only have the technology to visit space temporarily, and it is practical only to seek out places not too far from the Sun and engineer metastable minienvironments on our own scale. It is not now and may never be realistic to transform even small planets outside the habitable zone of a star into new island paradises.

There have been many champions of transforming a planet like Mars into a warmer, wetter place, but this is mind candy for science fiction buffs because so much energy (and time) would be involved. The cooling off of a greenhouse planet is tougher still because it requires dissipation of more heat than is being absorbed by an already hot object in a hot environment. This is like cooling off a roasting turkey without being able to turn off the broiler.

Some smart people, like the cosmologist Stephen Hawking, have argued that if we inhabit two planets, our odds of being killed off like the dinosaurs will decrease. This is a simple statistical fact given enough time, but the probability that even your great-great-grandchildren will die in such an event, thermonuclear war aside, is vanishingly small.

The two-home idea is also Carl Sagan's view on spacefaring, but Sagan meant more than that. He knew that Earth is the only planet in the Sun's habitable zone, that the challenges of the dual-home concept are enormous, and that the probability of a cosmic event wiping us out any time soon is miniscule. It is not just living on two planets that matters but the development of the technologically self-sufficient settlement. The capabilities for indigenous resource utilization are too far in the future to hedge against global catastrophe by using a moonbase supported from Earth. Spacefaring takes independent resources, not just stringing a tenuous web of lifelines around the inner Solar System. If the Moon actually had seas and an atmosphere, someone would have already figured out how to live there permanently.

You may have caught a glimpse of the erstwhile goblin of *how long* civilization will last and *how far* we will go. This is more speculative than forecasting climate change, but some think that the fossil record of extinction tells us how, as a random observer, to compute the probability of human longevity. The observer simply assumes she is alive at a point in time relative to the number of individuals that have ever lived or the number expected to live before the species dies out. These ideas were popularized by Carter, Gott, and others who compute the longevity of our species as well as estimate our ability to disperse throughout the galaxy (Gott 1993, 2007). Statistically, this argument is based on confidence intervals, but the calculations depend both on initial debatable assumptions and on threadbare extrapolations.

Gott assumes, for instance, that we are random, intelligent observers living in a nonprivileged location some 200,000 years into the evolution of our species and that there is a 95 percent chance we are seeing things between 2.5 and 97.5 percent of the way through the lifespan of our species. On this basis, he has set our longevity at less than 8 million years. More precisely, he figured we will be around another 5,000 to 7.8 million years.

Judge the value of this for yourself, but the argument is sensitive to the conditional use of time and to the definition of "intelligent observer." Since only recently has there been an observer to read the fossil record and perform these calculations, one could argue that only cognitive history is relevant to our longevity. If "mathematics" is 5,000 years old, this argument would predict 125 to 4,875 years left to our civilization and would thus argue *against* mathematics having a positive effect on our survival.

I find it awkward to argue against the use of brains and science to decrease risk and improve our chances. There is no way to know whether just being here or being here to know enough to perform the analysis is more important to survival. In other words, we are not random intelligent observers but instead special observers at a time and a place when it is possible to do calculations and interpret them as predictions. It would be unfortunate if we turned out to be too smart for our own good and annihilated ourselves. Either way, we are in no position to take the idea of a second cosmic home as a means to secure our survival too seriously yet.

The best course of action seems to be to learn as much as possible about our place in the universe while trying to avoid killing ourselves. The most famous instrument in cosmology, the Hubble Space Telescope (HST), looks into the past, mainly at large objects that are forever out of reach. Yet no one in their right mind has seriously objected to the $1.5 billion spent on it in 1990 because it changed our understanding of our universe (Delcanton 2009).

To keep Hubble going, NASA faced some legendary problems. Originally, four shuttle missions were scheduled to service it, but the fourth one to perform maintenance and install new instruments was cancelled in 2004 by then NASA Administrator Sean O'Keefe, who was rattled by the *Columbia* tragedy. In the clamor that ensued, a National Academy of Sciences panel recommended reinstatement of the mission, to which then NASA Administrator Michael Griffin agreed. By the end of 2006, HST began to shut down, and in January 2007, its Advanced Camera for Surveys (ACS) was out of business because of an electrical problem.

The fourth Hubble mission was finally performed in 2009 by the crew of STS-125. The astronauts installed a new wide-field camera and high-sensitivity spectrograph and repaired the ACS that had failed two years earlier. The old telescope became more powerful than ever and had the ability to see the universe back in the Reionization Era, at just 5 hundred million years after the Big Bang.

The Hubble flap was not just about money but about safety. Hubble orbits out there along with plenty of space junk, 350 miles (563 km) above the Earth and 130 miles above the ISS, and the orbiter risked taking real damage from fast-moving debris. And unlike the ISS, where shuttle astronauts could have stayed if they were unable to return to Earth on the orbiter, Hubble offers nothing. NASA put a standby shuttle on a launch pad should a rescue mission be necessary, but the mission went smoothly, and the crew returned safely.

The Hubble's heir is the $4.5 billion infrared James Webb Space Telescope (JWST), which has a mirror six times Hubble's area. The Webb will orbit beyond the Moon, where it will be more hazardous to repair than Hubble. Certain spacecraft can dock with it, but not the *Orion*. Moreover, independent cost estimates have figured the JWST at $6 billion, and it was delayed until 2015 (Mann 2011). Of course, the customary question of whether the payoff is worth $6 billion is unanswerable, particularly since it will peer into the unknown.

Large telescopes can be installed more or less permanently on the Moon, too. The negligible atmosphere and stable surface are advantages, and lunar telescopes can be operated from Earth and serviced from an outpost with incremental difficulty relative to their remote terrestrial counterparts. Clearly, lunar dust and radiation are disadvantages, but astrophysicists already take advantage of the clarity of extremely high and dry climates around the world, including the South Pole (Clery 2007).

Deep optical space telescopes will likely again revolutionize our understanding of the early universe, but to study very ancient structures, a cold telescope capable of weeks to months of Hubble-quality image integration is needed. A Lunar Liquid Mirror Telescope (LLMT) of twenty- to one-hundred-meter aperture diameter could observe objects in the infrared region a thousand times fainter than the JWST. This resolving power is important for seeing objects at the highest redshift, for instance, far older galaxies (Borra 2007).

Primary mirrors made by spinning a liquid in Earth's gravitational field have the potential for producing large, high-quality, low-cost optical telescopes. An undisturbed spinning liquid surface is almost perfectly smooth, and after being disturbed, gravity and inertia restore it to a parabolic shape. A liquid-mirror telescope on the Moon would avoid the need to transport large solid mirrors to the Moon. The main disadvantage, a narrow field of view, is marginally relevant if the cosmological focus is on hoary, deep-field objects.

For infrared LLMT, low-temperature, low-vapor-pressure, viscous liquids can be coated with a highly reflective metal surface to function as a telescope mirror. On the Moon, the coated liquid must have a low vapor pressure and a low freezing point. An ideal reflective metal film would have a high viscosity, low melting point, negligible vapor pressure, and would not sublimate in a vacuum. Liquids composed entirely of ionic salts have most of these characteristics; they are fluid at temperatures below 373 K, with no significant vapor pressure at room temperature. Ionic liquids are highly viscous, may be either hydrophilic or hydrophobic, and can be stably coated with silver (Borra 2007).

These examples illustrate the types of science payoffs that have been so tough to achieve on the ISS. Lunar exploration is well encapsulated at NASA, and the Moon's size and geology gives more traction to a science facility than on the ISS. Those who take exception to this step must account for the enhanced scientific value of combining human exploration with other kinds of space research. The potential practical spinoffs for clean energy, clean water, and new materials are much higher than for the ISS, which has been a laboratory mainly for the study of the human condition in microgravity and for EVA. Although these things are important to NASA, they only modestly benefit the rest of society. And no matter how far EVA capabilities and asteroid missions take us, the surface technologies for the exploration of Mars won't properly evolve without stopping on the Moon again first.

PART 2

A HOME AWAY FROM HOME

6. LIVING OFF THE LAND

The vision of a future pantheon of space explorers leading us across vast expanses of space to other worlds mirrors the major historical migrations of societies on Earth. This fantasy of the science fiction writer often deals with overcrowding and the tides of war, but migrations on Earth also fail because of the vicissitudes of geography, climate, and trade. The success or failure to disperse involves many variables, but ultimately, the comportment of the community accounts for all outcomes. Societies that collapse may not be distinguishable at first from durable ones with respect to adaptability and resourcefulness, failing only later through shortsightedness (Diamond 2005). The tipping points involve the population's ability to manage the resources in its environment, particularly at times of change. The rule of antiquity to some extent still applies: find good water and learn to live off the land as soon as possible.

The effective use of indigenous resources is a survival imperative, for should the new environment prove hostile, the population outgrow its resources, or the parent country become fickle, disaster is waiting in the wings. The longer the lifeline, the greater is the uncertainty and the less the permanency. Remote colonies under extreme stresses face dissolution regardless of the resourcefulness and hardiness of the people (Tainter 1988).

Historians spend a lot of time analyzing failed societies, often drawing fascinating parallels or projections to the modern social order. History offers less insight into space exploration; it is too young and has no terrestrial counterpart. On the other hand, it doesn't take a historian to see why we haven't colonized the Antarctic plateau or the continental shelf. There is no compelling reason to live on the ice or underwater. The same reasoning can be applied to the Moon and Mars.

Such nihilism would normally cause me to shut my computer off and go to bed. Still, we are in the spacefaring business, and there are lessons to be learned from failed colonies. Take the Norse: they found their way by sea to Greenland in the tenth century and established two settlements that thrived, but then they had vanished by the early fifteenth century. The colonists died of cold and starvation—they were no longer able to subsist in a marginal environment for reasons that are now well understood.

The fate of the Greenland Norse seems predictable today, and even with climate change, modern utilities, and rapid transportation, Greenland, the world's largest island, has a population of only 57,000, mostly Inuit, people. They live on the southwestern edge of an island a quarter the size of the continental United States, at a density of one person per fifteen square miles.

Five hundred years ago, the Norse had no way of knowing that the island's tiny arable land area, its short growing season, and deforestation and soil erosion were terminal problems, compounded by their great distance from home and recurrent conflict with the Inuit. When the Little Ice Age emerged, Greenland's climate changed, and the Norse could not last. The colonies were linked to Scandinavia by thin threads, and over some four hundred years, they slowly strangled. They infrequently received critical supplies such as iron and lumber, and they had little to trade, primarily walrus ivory and furs. When an abrupt climate change further shortened the growing season, clogging the trade routes with ice, they were finished.

In order for the history of Greenland not to adumbrate the fate of lunar or Mars bases, which face far greater environmental extremes, there must be another way. A Mars outpost would be at a far greater risk of extinction than the Norse were, unless the technologies can be developed for generating power and harvesting resources efficiently. The time-honored strategy of living off the land in the Space Age is often called ISRU, but I find the acronym ISLE—

Indigenous Support of Life and Environment—drives the point home best. ISLE involves critical surface technologies that currently exist only on paper.

ENERGY AND EFFICIENCY

ISLE is necessary when distance and orbital mechanics make cargo missions from Earth unworkable. The distances for Mars and beyond are too great to depend on the Earth for cargo, and ultimately, ISLE determines the feasibility of human exploration of the Solar System. A sound ISLE strategy first requires energy and energy-efficient technologies, which will evolve from prototype to praxis beginning on the lunar surface.

ISLE begins with *power*, the rate of energy production and/or energy use. Energy is transferred to do work, and work is the rate at which energy is used. Energy divided by time is power, and we will use the SI unit for power, the watt (W), which is one joule per second. One joule (J) is roughly the energy lost when Newton's apple fell one meter.

Electrical power is the modern equivalent of fire, without which we would still be confined to Earth's most temperate zones. As subtropical creatures, we adapt poorly to cold, and our ancestral comfort zone stands in the narrow band of latitudes thirty-five degrees north and south of the equator. Our utilities are based on technologies with huge capacities to generate and store electricity for use on demand. We are accustomed to electricity, but we evolved without it, and if it disappeared tomorrow, most of us would survive. Earth is fed by the Sun, and we are adapted to our world's temperature. We use extra power to make moderate seasonal adjustments in our immediate surroundings in order to live more comfortably.

Every spacecraft must have the power to operate its life-support systems in a sustained mode, and the rocket's burst is nothing compared with the continuous power needed to supply a safe and comfortable habitat in space. An offworld station or base, like a modern city, must generate and store power for use on demand.

On the other hand, robotic or "unmanned" probes simply need a bit of power here and there to make course corrections, run their instruments, and report home. As we saw from *New Horizons*, the power requirements are modest if onboard instruments operate only intermittently and can safely "standby" at low temperatures. You and I have no standby mode.

The technologies for generating electrical power on the Moon or Mars are not ready yet, although the energy mavens lean toward combinations of solar and atomic power. Those technologies are safe and reliable, notwithstanding the dreadful accidents at Chernobyl and Fukushima. Modern radionuclide thermoelectric devices that power robotic space missions, such as the *New Horizons* and the Mars *Curiosity*, are too small to support people, but nuclear power, which is capable of doing so, gets handled by NASA like a hot potato.

Solar power does work on Mars, but the Martian winter and Martian dust are tough on solar arrays. Solar power is not practical in the outer Solar System because collecting sunlight on small surfaces is inefficient. Based on the inverse square law, solar collectors—even highly efficient ones with concentrators— will work only within about ten AU of the Sun. There may be a few alternatives, such as wind power, in places with meaningful atmospheres, like Mars.

The power requirements of human spacecraft are high, and power is needed continuously and on demand. For LEO, as on the ISS, power is stored in batteries charged by solar panels. The Moon and Mars will benefit from the use of local materials to generate energy and regenerate consumables to minimize dependence on resupply missions.

Manmade systems, like living organisms, can draw on stored energy for a while, but sooner or later the stores must be replenished. We eat regularly in order to perform work, and those who cannot eat will die from starvation within weeks. The harder we work, the more energy we use and the more quickly our reserves can be depleted. This is the case for all natural and artificial working systems.

Some natural cycles use energy to generate heat or light that can be utilized by other parts of the cycle. This may improve a cycle's efficiency, but the second law of thermodynamics is unbreakable; it prohibits perpetual-motion machines. The tendency for disorder (as entropy) requires that external energy be applied at some step and for some duration to maintain any "self-sustaining" cycle. Moreover, the conversion of energy from one form to another, for instance, sunlight to electricity, is associated with the loss of energy as heat. The second law prohibits energy transformations having *thermodynamic efficiencies* greater than or equal to one (Rao et al. 2004).

Thermodynamic efficiency is the overall efficiency of any work cycle according to the ratio of work done to the amount of heat generated by the work. For instance, the thermodynamic efficiency of an internal combustion engine, depending on the type of fuel and the operating temperature, is about

50 percent. This means that half of the energy released by burning fuel escapes from the engine as heat without producing mechanical work. In a set of linked systems, the product of the efficiencies of each component determines the overall efficiency.

Self-sustaining systems on spacecraft behave similarly, but the calculation of efficiency is not simple. On a spacecraft, systems that regulate the atmosphere, temperature, and water and that produce and process food and remove and recycle waste require more power than it takes to do the work. And extra power is required for master functions like integration. This must come from a power supply, batteries, for instance, that can handle several systems at peak loads simultaneously. The excess heat must be dissipated or used for other purposes.

In spacecraft, mission parameters such as crew size, the type of work to be done, and duration set the life-support specifications. These specifications encompass the mass and volume of O_2, H_2O, food, and other essential supplies, including their containers, as well as the physical equipment for collecting, storing, and recycling waste. These systems are configured mostly inside a hard shell, along with a shielded cabin of sufficient volume for the needs of the crew. This hard shell is wired to a power storage system.

The principles of life support were originally worked out for submarines during World War I and have clearly passed the test of time. For long-term subsea power, the small atomic reactor was the key, but in space, sunlight is still king because solar technology, although inefficient, is clean, safe, and lightweight. NASA has favored photovoltaic technologies, which utilizes silicon cells to convert sunlight directly to electricity. Large numbers of solar cells assembled into arrays are used to boost power levels and to charge batteries.

The ISS is powered by 110-foot solar wings consisting of solar cell "blankets," one on either side of a telescoping mast that extends and retracts the wing. The mast rotates on gimbals so the wing always faces the sun. A pair of wings is a "module." Four modules generate eighty-four kilowatts (kW) of power and cover about three-quarters of an acre.

The ISS spends about a third of every orbit in Earth's shadow, so the electrical system provides continuous power via rechargeable, nickel-hydrogen batteries that are recharged during the daylight part of the orbit. The process of collecting sunlight, converting it to electricity, and distributing it throughout the station, apart from its low efficiency, is considered quite satisfactory by electrical engineers. Heat buildup can damage the electrical systems, and surplus heat is dissipated with radiators that are aligned away and shaded from the Sun.

The levels of power generated by your local power company are too high for routine household use, and the same is true for the ISS. On the ISS, the power output is stabilized and shunted to batteries or to switching units that route it to transformers that step it down to 124 volts. The original hardware would generate roughly as much power as used by fifty homes.

The solar cells on the ISS use older technology that converts about 14 percent of the light that hits them into electricity. Newer technologies to convert light from multiple parts of the spectrum into electricity (such as multibandgap cells) are about twice as efficient. These devices perform well in the inner Solar System but may never be adequate in the outer system.

As the distance from the Sun increases, the intensity of sunlight falls by the inverse square law ($1/r^2$), where r is the distance (see figure 6.1). This law means that a one-square-meter (1 m^2) array that produces 500 watts on Earth (or the Moon), at 1 AU, would need 2.25 m^2 at Mars and 25 m^2 at Jupiter to generate the same amount of electricity. At Neptune, the array would need to cover roughly 900 m^2 (3,000 square feet). This array would collect enough sunlight to operate a kitchen microwave oven.

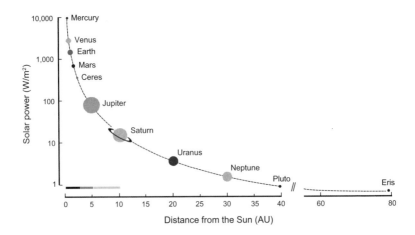

FIGURE 6.1. THE DECLINE IN SOLAR IRRADIANCE WITH INCREASING DISTANCE FROM THE SUN IN OUR SOLAR SYSTEM.
The steepness of the curve reflects the inverse square law. The bars at the lower left show, respectively, the farthest practical demonstration of solar power in space, the limits of current technology, and its possible extension with collector technologies under development. Most of the outer Solar System cannot be reached using solar power.

Improvements in power generation for a Mars mission will involve high-efficiency solar cells, but more exotic proposals, like a tiny fission reactor on the surface or beaming laser energy to the spacecraft, are under serious consideration. Today's solar technologies are limited to about five AU, or about 4 percent of Earth's sunlight. They would work on Ceres (2.7 AU) and perhaps at Jupiter (5.2 AU), but even with better concentrators and boosters, solar power near Saturn (10 AU), which receives 1.25 percent of Earth's sunlight, may not be feasible.

The record for operation of a solar-powered spacecraft was set in 2002 by NASA's *Stardust* probe en route to the comet Wild 2. It reached 2.72 AU, about midway between the Sun and Jupiter. *Stardust's* solar arrays performed better than expected, possibly because its photovoltaic cells were more efficient in the cold of deep space. In any case, *Stardust* confirmed that solar collectors will operate at least as far away as Ceres.

The *Deep Space 1* (DS1) probe in 1998 tested advanced spacecraft and instrument technologies for use on interplanetary missions. DS1 was powered by xenon-ion propulsion, or an ion engine similar to those used for satellite stationkeeping. The DS1 ion drive accumulated more operating time in space than any earlier system. DS1 also put the first modular solar concentrator into space, called SCARLET-II (Solar Concentrator Arrays with Refractive Linear Element Technology), which achieved a high performance with 720 lenses that focused sunlight onto 3,600 high-efficiency solar cells. This provided DS1 with 2.5 kW at a power-to-weight ratio of about 50 W/kg, the most weight-efficient solar array ever flown. A lightweight, prototype Fresnel lens that focuses sunlight onto narrow strips of photovoltaic cells has put out 300 W/m^2, several times better than a planar array. Fresnel lenses are arched microscopic prisms that enable up to 90 percent of the incident light to be focused onto solar cells.

By concentrating light onto a spot, the amount of photovoltaic material can save 95 percent of the cost of solar cells, often the most expensive part of an array and an expensive part of a spacecraft. Solar cells should eventually become thin lightweight films that could be pressed out to acres in size. Such "sails" must be exceedingly strong and durable to limit degradation by cosmic radiation and micrometeoroids. Even as materials and concentrators continue to improve, allowing for more distant missions, the point of diminishing returns will never be far away.

In the sea, sunlight does not penetrate below about three hundred feet, but the power problem for submarines was solved more than fifty years ago with the atomic reactor. Reactor technology, despite some unearned notoriety, has rapidly advanced in terms of efficiency and safety. Small pressurized-water reactors on submarines provide as much as 190 megawatts. The reactor cores are long lived, requiring refueling every fifteen to twenty years. Newer reactors on modern aircraft carriers can operate for twenty-five years or more.

There is ongoing interest in small reactors for use in remote locations, and although fissionable material is precious and prone to misappropriation, the options for packaging and safely disposing of it in space are already reasonably far along. America has owned several small, land-based nuclear power plants, such as the little reactor that operated at McMurdo Sound in Antarctica between 1962 and 1972, generating some seventy-eight megawatt-hours of electricity. Small reactors have operated successfully for thirty-five years, and reactors with hundred-year lifetimes may be available before the end of the twenty-first century.

This leads us to an analogy between Antarctic and lunar exploration. Since 1956 and the installation of the original U.S. Navy station, Americans have occupied the geographic South Pole continuously. Situated on Antarctica's two-mile-thick ice sheet, the Amundsen-Scott South Pole Station is some 850 nautical miles south of McMurdo Station, at an elevation of 9,306 feet (2,835 meters). A geodesic dome built in 1975 accommodated eighteen people in the winter and thirty-six in the summer. The third South Pole Station, dedicated in 2008, supports an expanded NSF Antarctic Research Program. The $150 million facility can support fifty people in the winter and 150 in the summer.

The operation of the station is expensive, but the research is important, particularly to atmospheric physics and meteorology. It is dark for six months, between February and October, and every supply crate (except water and O_2) is shipped in and every scrap of waste shipped out during the austral summer. The station contains a small NASA plant-growth facility and a power plant that burns cold-weather JP-8 fuel, which generates up to one megawatt. The power plant operates continuously to prevent the facility (and people) from freezing, and it has three levels of backup. The backup options are limited to wind turbines and solar power, the latter only during the summer. The likeness to a moonbase is striking, continuously generating power for heat, but on the Moon it would be called on to produce a breathable atmosphere as well—and without burning fossil fuel.

ECOLOGICAL FOOTPRINTS

The South Pole Station conjures up visions of living in cold confinement for months at a time on the Moon or Mars. I made a point of energy and efficiency because air, water, heating, and the recycling of resources use lots of power. A related area of concern for NASA is the lack of information on how a full community in space environments functions in isolation.

The science of organisms in their environments, or *ecology*, was defined in 1866 by Ernst Haeckel. Human ecology, compared with terrestrial, botanical, or marine ecology, is unique because the ecologist hopefully has her own human experience: she is a topic of the discipline. This is counter to the dualism in science, evident since Descartes, that the scientist is in some way different from that being studied, Nature.

Even the second law of thermodynamics sets life apart—physical systems run down or become more disordered, but life temporarily "runs up" by creating local order. The tendency for physical processes to run down is contrary to the directed behavior of life. Rivers run downhill, but salmon swim uphill to breed; gravity causes heavier-than-air bodies to fall to the ground, but bats, birds, and bugs fly. The engineer epitomizes the principle of counteracting the second law through the process of invention and design.

The emergence of aerobic life is another case in point; when Earth developed an O_2 atmosphere by oxygenic photosynthesis, the evolution of aerobic life leapt forward. The change exemplifies evolution, and the geological record speaks volumes on the importance of change. Yet critical changes or tipping points can also lead to mass extinctions, some of which are now being set off by us, too.

Living systems are spontaneously organizing and self-replicating; they maintain order by breaking down energy-rich compounds from the environment and releasing them in an energy-depleted state. Such *autocatakinetic* processes are not limited to living systems (Swenson 1997). Our atmosphere, for instance, is punctuated by highly organized storms, such as tornadoes and typhoons, and similar storms on other planets indicate that spontaneous organization is a property of matter. This propensity to create order spontaneously accompanies large potential energy differences between sources and sinks.

This line of thinking suggests that large ecosystems and the large differences between them and their energy sources might be working to our advantage. A system with a high efficiency may have a limited capacity to evolve, by not having enough energy to change how it operates. In other words, size, surplus energy, and room to improve efficiency may favor sustainability. Apart from economies of scale, these ideas have not been seriously tested, but even after billions of years of terrestrial evolution, living systems tend to operate at fairly low efficiencies.

Ecologists vary in their views of behavioral adaptation and its relationship to biological evolution, and many human ecologists think that cultural evolution progresses differently than biological evolution. Since we cannot adapt physiologically, space exploration is fundamentally a problem of behavioral evolution. The analysis of human communities for ecological sustainability involves an element of behavioral forecasting, but like long-range forecasting of the weather, the predictions, especially with respect to space, have huge uncertainties.

On Earth, the pool of resources available to any population is finite. Given unrestricted population growth and time, some resource will eventually become limited (Pianka 1994). This calls Thomas Malthus to mind, whose theory of inevitable global famine has been held at bay by technological improvements that produce and deliver food to our expanding population. Moreover, large communities tend to behave as though no one is responsible for the resources we hold in common. This "tragedy of the commons," popularized by Garrett Hardin, is well appreciated in ecology and economics. Visible and troubling examples would be the beehive of junk satellites in low Earth orbit, the great Pacific plastic patch, and the rising atmospheric CO_2 emissions from burning fossil fuels.

Ecologists have long watched our comings and goings for signs of sustainability, and they entertain themselves by calculating the hectares of photosynthetic capacity that are needed to support one person in one community in comparison with other communities. Earth's useable land area is finite and has a maximum capacity, which if exceeded, either by overpopulation or by living beyond our means, degrades societal structure. This is the essence of the *carbon* or *ecological footprint*. To Americans, carbon footprints are contentious because we have little interest in giving up lifestyle resources to the rest of the world. For fifty years, we have lived at roughly four times the sustainable footprint. We are happy for others to "catch up"—as long as we get to keep our two SUVs, four TVs, and 4G smartphones.

In space, the size of the human footprint gets larger, not smaller. This is true for the ISS, for the Moon, and for Mars. This is the technological imperative; Spaceship Earth suits us best, and living in space requires techno-logical *intensification*. As a result, communities in space offer no escape from population growth on Earth (Daly 1993). Even on our own planet, drastic agricultural intensification to support local over-population has already been the root cause of ecological disasters.

ON NEVER RUNNING OUT OF AIR

Since the amount of O_2 on a spacecraft is finite, and because gases inexora-bly seep to the outside, the losses must be accounted for on long missions. Each time an airlock on the ISS is opened for EVA, a little O_2, N_2, CO_2, and water vapor are drawn into space. And gases do escape from pinholes here and there, and that adds up over time. O_2 is also consumed or adsorbed by spacecraft structural materials.

The ability to supply O_2 in space depends on the stock of three mole-cules: O_2, H_2 and H_2O. In order to illustrate why, we'll start by tracking water through a life-support system. Water is a nice example because it passes un-adulterated through the body, and it can be collected, purified, and reused at a low cost and a high efficiency relative to other resources. Water is not only for drinking; it is also used for generating O_2 by electrolysis. Sunlight can be used directly for this process, too (Wrighton et al. 1975), and water and electricity can also be made by electrolysis by combining hydrogen (H_2) and O_2. Both processes require energy. H_2 is readily available; it is, after all, the universe's most abundant element, and fuel cells combine H_2 with O_2 to generate H_2O with the release of energy.

For electrolysis, the energy needed to split the H_2O molecule is approxi-mately 237 kJ per mole (one mole of H_2O is eighteen grams). A fuel cell combines O_2 and H_2 to produce H_2O and electricity, with 237 kJ recovered as electrical energy and 48.7 kJ dumped out as heat. At first blush, the latter seems more attractive, except that energy was expended to collect and com-press O_2 and H_2 and that kinetic energy is there in the gases before combina-tion. On Earth, O_2 is "free," and energy is added to prepare H_2. However, O_2 is not free on the Moon. It must be transported from Earth or extracted from the soil.

The Moon has O_2 and H_2O, but the challenge is to generate the energy and design equipment to harvest them in amounts sufficient to live safely. The recovery of O_2 and H_2O on the Moon has been the subject of a dizzying number of proposals, many based on speculative technology, but luckily the problem can be illustrated independently of specific technology. We can define it through *limits* without actually choosing a technology. Moreover, the finite supply of O_2 and H_2O highlights a classical dichotomy: if one resource depends on the other, it is impossible to optimize both. Therefore, an optimal strategy would use independent sources of O_2 and H_2O.

Let's examine H_2O recovery on the Moon. Assume the moonbase has an initial reserve and that harvestable water is nearby. In other words, Allen Stern won't be shooting blocks of ice up from the Earth. We need one more bit of information for the H_2O spreadsheet: the efficiency of water recycling in our system.

Today's water recovery technologies, like reverse osmosis and distillation, leave behind nasty slurries and brines, and technologies that recover water efficiently from brines and slurries are of great interest because they increase the conservation of mass, or *mass closure*, of life-support systems. The residual slurries of dissolved solids and organic material are 15 to 20 percent water by weight, which means today's best water recovery technologies are about 80 percent efficient, although on paper, there are systems that may achieve 95 percent efficiency.

If you had visited a town reservoir during the drought in the southeastern United States in 2007–2008, the manager could have told you the number of days of water left. She would also not let the reservoir fall below some minimum, say, a thirty-day supply. To prevent this, she had to know two things: the volume of recycled water and the volume of water to buy from another town to compensate for permanent losses. Eventually, the town council would also want to know the price of water and how much energy the processes use in order to pay for it.

The amount of water to add back to the system is simply determined by the amount that permanently escapes the system each month. Since we need a constant reserve, the rate of water use can be ignored as long as water is recycled as quickly as it is used. We must harvest and add the amount of water lost permanently to space to our reservoir each month. Suppose a recycling plant on the Moon dissipates seventeen liters of water into space and dumps thirty-three liters of water as brine each day. Every month, you must add 1,500 liters of water to the reservoir. Notice that this number is in absolute liters, and the recycling efficiency and the energy cost are hidden in the volume of water.

If the overall efficiency was 80 percent but is improved to 96 percent, the number of liters lost per month falls to 1,250, and 250 fewer liters of water must be brought in from the outside. In other words, three factors govern the outpost's water economy: (1) the reserve volume, (2) the efficiency of water recycling, and (3) the availability of new water. A high recycling efficiency minimizes the irreversible loss of water that must be made up in other ways. However, the energy cost of efficient recycling is hidden, and it could be more expensive than simply finding new water. If water is abundant, it may be cheaper just to find more. This is like recycling aluminum cans. When the cost of aluminum rose because bauxite became scarce, aluminum recycling began to pay off.

The point is that the cost of obtaining a resource (water) and the cost of recycling it is an important tradeoff. If we use water both for drinking and for making O_2, the water losses go up, and we will need a closer analysis of recycling and harvesting. If water is scarce near the base but lunar oxides are common, it may be cheaper to heat rock for O_2 than to split water. It may also be more economical to combine some of the O_2 with H_2 to make water.

I have assumed that the power for all this harvesting and recycling will be available, but this is not the same as having unlimited power, only that our system operates like the Earth's biosphere, where the energy supply exceeds energy demand. If there is too little energy for harvesting and recycling, energy itself becomes a limiting resource.

Water sits just below O_2 with respect to life support. The generation of energy for the great bulk of homeostatic and physiological functions depends on O_2. The loss of O_2 interrupts the functions of the heart and brain, producing almost immediate loss of consciousness. In eight minutes, irreversible brain damage commences from hypoxia. We do not store O_2 in our tissues, and thus it must be supplied continuously to our lungs as fresh air. When the O_2 content of the blood falls below normal, it is called *hypoxemia*. Hypoxemia compromises survival because tissue hypoxia soon follows, which wrecks cellular function.

This is not the design I would have chosen for the human being; I much prefer that of the dolphin or the whale, which can hold their breath for an astoundingly long time. In any case, we are more susceptible to hypoxia than cetaceans because they store extra O_2 in their bodies. Some whales stop breathing for an hour while diving, relying on O_2 carried in their lungs and muscles. But we require the nearly continuous movement of O_2 from the atmosphere into our lungs, and a well-trained, resting breath-hold diver can last about ten minutes. Indeed, a diver that first breathes O_2 instead of air can extend his breath-hold time by about 50 percent.

There are four ways to run out of O_2, and they are all equally deadly. Three were first recognized by a distinguished physiologist, Joseph Barcroft, in the early twentieth century (Barcroft 1914). He distinguished among a low O_2 concentration reaching the blood because of high altitude or impaired gas exchange by the lung (*hypoxic hypoxia*), too few red blood cells in the circulation (*anemic hypoxia*), and too little blood flow to a tissue (*stagnant hypoxia*), now called *ischemia*. A fourth category was added later, in which the utilization of O_2 by the cells is impaired; this is *cytotoxic hypoxia*. A cardiopulmonary arrest is the most extreme form of global ischemia, but our primary concern here is hypoxia arising from low O_2 concentrations in the atmosphere.

In the atmosphere, barometric pressure falls with altitude (shown in figure 4.1). Although the percentage of O_2 remains constant, the total number of O_2 molecules in any volume of air falls. Recall that barometric pressure at sea level is 760 mmHg and that air contains 20.9 percent O_2; therefore the *oxygen partial pressure* (PO_2) is 159 mmHg. Atop Mt. Everest, where the barometric pressure is only 253 mmHg, the inspired PO_2 is about 53 mmHg. This is just enough O_2 to survive after being fully acclimatized, and at the summit, arterial PO_2 values near the limit of human tolerance are measurable (Grocott et al. 2009).

In breathing, air in the distal or gas-exchange regions of the lung is brought into contact with hemoglobin-containing red blood cells, which bind O_2, carry it to the tissues, and release it for cell respiration. Thus, in air-breathing vertebrates, the external and internal milieus of the body are connected via the lungs (salamanders, frogs, and perhaps some tiny mammals also absorb O_2 through the skin).

We live at the precipice of hypoxia because O_2 is consumed irreversibly by the process of respiration, and respiration is irreversible because O_2 is reduced to water and excreted. In the process, sugars (mainly), fats, and proteins are broken down, heat is generated, and chemical energy is trapped by the addition of inorganic phosphate (P_i) to adenosine diphosphate (ADP) to make adenosine triphosphate (ATP). Without ATP, the cell quickly ceases to function. The elementary chemical equation for respiration is:

$$C_6H_{12}O_6 \text{ (glucose)} + 6O_2 + P_i + ADP \rightarrow 6CO_2 + 6H_2O + ATP$$

The equation indicates not just that O_2 and glucose (or other nutrients) are taken from the environment but that O_2 is excreted as CO_2 and water. The complete oxidation of a glucose molecule releases 686 kcal of energy (1 kcal = 4.18 kJ), and the cell can trap about 40 percent of this energy as ATP. It also takes four electrons (plus four protons) to fully reduce O_2 to $2H_2O$.

Energy extraction from hydrocarbons with the help of O_2 works because of the molecule's unique chemistry. Yet this gas was not a factor in the origination and early evolution of life on Earth, and our atmosphere originally contained little or no O_2 at all (Lane 2003). O_2 was toxic to many early single-celled organisms that lacked defenses against its tendency to strip electrons from macromolecules by *chemical oxidation*. Such O_2-intolerant organisms, or anaerobes, are still abundant on the Earth today, especially deep in the soil.

Chemists measure the potential for O_2 to capture electrons with an electrochemical cell called a *standard cell*. Chemically, O_2 is highly electronegative, $+0.82$ volts relative to hydrogen. This means O_2 is capable of oxidizing most hydrocarbon compounds. Oxidation slowly dissipates chemical energy introduced by other processes, such as photosynthesis. Photosynthesis traps the energy of sunlight, while hydrocarbon oxidation releases energy.

This energy release is accelerated by catalysts, for instance by metals like iron, which promote the rapid transfer of electrons to O_2 from less electronegative donors. Iron is present in the respiratory centers of most cells, but at a cost. O_2 catalysis can generate free radicals, or more properly reactive oxygen species (ROS), which freely oxidize biological macromolecules. The discoverer of O_2, Joseph Priestly, pointed out that we breathe "air as good as we deserve"; that is, the O_2 concentration in the air is optimized for our form of life. Very high O_2 concentrations, or *hyperoxia*, can damage the lung, the brain, and the eye, and we have a modest range of O_2 tolerance because of our antioxidant defenses. In 1954, Rebecca Gerschman and Daniel Gilbert also discovered that cellular damage by O_2 shares a common mechanism with radiation and involves ROS generation (Gerschman et al. 1954).

The geological record indicates that about 2.4 billion years ago the O_2 partial pressure (PO_2) in the atmosphere increased rather sharply—by a factor of about 10^5—to its present level (Kasting 2006, Anbar et al. 2007). This transition is commonly called the Great Oxidation Event, or GOE. The thinking is that the GOE reflects the evolution of oxygenic photosynthesis by cyanobacteria (blue-green algae), which began to flourish in expansive mats during the Paleoproterozoic Era.

Indeed, oxygenic photosynthesis may predate the GOE by two hundred million years. The rise of the atmospheric O_2 may have been delayed, for instance, because the chemical reactivity of O_2 changes as its levels fluctuate, but the length of the delay suggests that other factors are involved. There may have been geological sinks—processes that remove O_2 from the atmosphere, such as the eructation of reduced gases from the Earth's mantle or the discharge of ferrous iron (Fe^{2+}) from hydrothermal vents. Thus, despite the burst of oxygenic photosynthesis, O_2 was initially removed as fast as it was being introduced. Once this scrubbing process had dissipated, the atmospheric O_2 concentration rose significantly.

Other biological factors may also have been needed for the GOE, such as the trapping of bacteria in marine sediments or carbon burial. This leaves O_2 behind in the atmosphere because organic material that precipitates in deep ocean sediments cannot be oxidized. But the predicted changes in the ratio of organic carbon to carbonate (CO_3^{-2}) for relevant geological strata have not been found. Whatever the case, oxygenic photosynthesis won out, and it is the only important source of molecular O_2 on our planet today.

The geological record also indicates that oxygenic photosynthesis was not the first form of photosynthesis (Battistuzzi et al. 2004). Perhaps half a billion years before oxygenic photosynthesis appeared, a type of photosynthesis appeared in which O_2 is not released, and some bacteria still utilize it. Moreover, the source of electrons is not water but other inorganic compounds, including those of the iron cycle, the sulfur cycle (sulfide), and the nitrogen cycle (nitrite) (Griffin et al. 2007).

In the evolution of oxygenic photosynthesis, two types of photoelectric devices were brought together, presumably by gene transfer, to make an efficient electrochemical cell that splits water into hydrogen and oxygen (Allen and Martin 2007). These devices—termed photosystems I and II—operate together in green plants in concert with chlorophyll and the hundred or so proteins of the chloroplast to capture energy from sunlight.

Photosystem II captures four photons, using them to remove two electrons from each of two H_2O molecules, and it releases an O_2 molecule into the environment as a product. In modern plants, photosystem II operates only in the presence of photosystem I, which takes the four electrons generated by system II and four additional photons to reduce a chemical nucleotide called $NADP^+$ to NADPH. Eventually, these electrons are transferred to CO_2 to provide the building blocks for sugars and other hydrocarbons.

One of the scientific challenges of understanding oxygenic photosynthesis has been to decipher the chemical evolution of the water-splitting complex of photosystem II. This complex contains a metal cluster, manganese-calcium (Mn_4Ca), where it interacts with the H_2O molecule and interfaces with the environment. The manganese cluster sequentially releases electrons to photosystem II proteins under the influence of sunlight, but only when the cluster is four electrons short does it split two water molecules to replace them. The capacity to do this may be unique to manganese, and the water-splitting chemistry of the Mn_4Ca cluster has long been sought in the quest to exploit sunlight as a clean, renewable source of power.

Once our atmosphere converted from a reducing to an oxidizing one, Earth rapidly evolved nonphotosynthetic organisms that captured and reduced O_2 to survive (Anbar et al. 2007). They took advantage of the chemical reactivity of O_2 in the presence of transition metals, but this added the requirement for antioxidant defenses. The spatial localization of these defenses allows O_2 to be transported safely some distance (micrometers) in tissues before it is used. Once O_2 could be transported and distributed on a scale of centimeters to meters, the evolution of higher organisms received a boost from a higher ratio of body mass to surface area and greater specialization.

WATER AND FOOD

On Earth, where O_2 is not limited, H_2O is *the exigent survival requirement*. A healthy person can live weeks without food but only, depending on the temperature and how much she sweats, four to seven days without water. Our bodies are roughly 60 percent water, but every day, liters of water evaporate from our skin and our lungs and are excreted by our kidneys. This water must be replenished to avoid dehydration.

In the desert, you can die in a day of dehydration because extra water is not stored, and we need it to keep cool. Sweat cools the skin as it evaporates because the latent heat of vaporization carries excess body heat away. The water is lost from the blood plasma, the spaces between cells, and later even from cells themselves. We get thirsty when the salt concentration (or osmolality) of our plasma rises too much, and we must drink regularly to replace the water that evaporates. And the kidneys must lose about one liter of water per day.

Compared with dehydration, starvation is slow because we have nutritional reservoirs in body fat and muscle protein. Even a thin person is 40 percent muscle by mass, and about half of this can be wasted in the interest of survival (Collins 1996). This means the average person can live for several weeks without food.

In spaceflight, astronauts carry all the H_2O and food aloft that they need, or they have it resupplied regularly from the ground. On the ISS, H_2O is recycled, and food comes prepackaged. There are significant water losses both inside the station and into space, and one cannot tote water and food long distances in space. "Packing everything in" may get us to Mars once or twice, but it is not a permanent solution to the problem of exploring other planets.

The limits of transporting H_2O and food obviously depends on the efficiency and cost of space transportation technology, but many NASA scientists think packing tons of food for a mission to Mars may not be smart. As space transportation and food technology improve, this issue should be easy to resolve. At some point beyond Mars, distance and time supervene, and water conservation and food production will become the realities.

Like O_2 and H_2O, the amount of food loaded onto a spacecraft will depend on the efficiency of its food recycling system. The goal is to recycle as much as possible so that energy but little or no mass is added. The original food mass is eaten, digested, excreted, collected, and regenerated as efficiently as possible. Such systems are built on simpler co-cycles and subcycles, which depend on each other in important ways.

You saw earlier that the water cycle is easy to maintain because we do not break down H_2O chemically. H_2O is a physiological solute; it is absorbed, excreted, secreted, evaporated, or transpired, unadulterated, in the water cycle. NASA's water cycle numbers are 3.5 liters/day for the body and thirty liters/day for hygiene.

The water cycle is coupled to the carbon cycle, sometimes called the food cycle, which is more complex. These two cycles operate for all animals on Earth. Chemical reactions convert carbon substrates to products for specific energy-requiring functions, like respiration, while plants replenish the substrates by oxygenic photosynthesis. The carbon cycle connects O_2 and H_2O recycling from respiring animals to photosynthetic plants (figure 6.2).

An effective food cycle is a basic requirement for deep space exploration, and the basic elements are most easily seen by working with only O_2 and CO_2. In practice, the water and complete food cycles are critical, but we can understand how these critical resources become limiting in space exploration

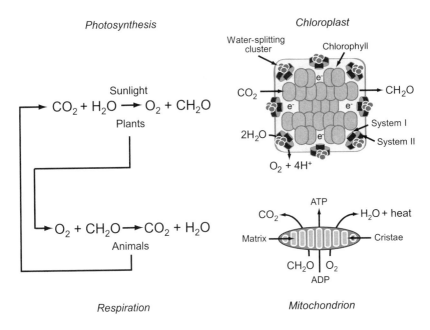

FIGURE 6.2. THE NATURAL CARBON CYCLE OF PLANTS AND ANIMALS ON EARTH.
Plants perform oxygenic photosynthesis in their chloroplasts, which require sunlight, CO_2, and water to produce sugars and other carbon substrates, while animals use carbon substrates, reduce O_2 to water, and produce CO_2 in their mitochondria.

with a simpler model. We need some food science, too, especially an estimate of our food requirements.

The history of food science at NASA is fascinating, and in aggregate, the space program has come a long way since the days of squeezing Spam out of tubes like toothpaste. Now, prepackaged menus consisting of three complete meals a day are sent up. They can be rehydrated and heated, and they meet the minimum recommended daily allowances (RDA) of vitamins, minerals, and protein (Smith et al. 2001). For the shuttle, these menus weighed about 1.7 kg, of which 0.5 kg was in the package itself; so the weight ratio of food to package was about three.

The nutritional value of space food is high, but the U.S. crew on the ISS must participate in regular nutritional evaluations that track dietary intake and body composition and the status of protein, bone, iron, minerals, vitamins, and antioxidants. People today tend to listen to popular advice and worry more about the calories, fats, and carbohydrate in their diets than about vitamins, antioxidants, and minerals. On future lunar and planetary missions, deficiencies in vitamins and other micronutrients that the body cannot synthesize will be an important problem to resolve. Nutritional deficiencies, like the dreaded scurvy of past sea voyages, obviously must be avoided.

On the ISS, food packages contain a barcode that the crew member scans to keep track of the daily calorie and nutrient intake. Food experts use the data to understand nutritional factors that help limit the adverse consequences of spaceflight, such as bone loss, shortages of vitamins or minerals, and damage to the body from radiation. The purpose of this monitoring is to develop optimal dietary strategies for human health in space.

The caloric requirements of astronauts are determined by two National Research Council formulas for basal energy expenditure (BEE in kilocalories; kcal). For women,

$$BEE = 655 + (9.6 \times weight) + (1.7 \times height) - (4.7 \times age)$$

For men,

$$BEE = 66 + (13.7 \times weight) + (5 \times height) - (6.8 \times age)$$

In these formulas, weight is in kilograms (kg), height in centimeters (cm), and age in years. Sample calculations indicate that a 170 cm tall, 70 kg woman, age 40, needs roughly 1,430 kcal, while a 180 cm, 80 kg man of the same age needs roughly 1,790 kcal a day. Taking into account extra calories for activities and exercise, typical astronauts require 2,300 to 3,100 kcal per day.

Astronauts generally eat less in space than they do on Earth, despite close attention to dietary preferences and palatability. Weight loss and loss of muscle mass and bone density are inevitable in microgravity. Dietary calcium absorption also decreases in part because of lower vitamin D levels because of the lack of sunlight. The bones become thinner and more brittle, especially the weight-bearing, antigravity parts of the skeleton. Thus, astronauts are encouraged to eat and are given calcium and vitamin D supplements.

Bone loss in space increases the excretion of calcium in the urine, which creates a propensity to form kidney stones. Calcium excretion may be accentuated by the relatively high protein intake of astronauts—eighty-five to ninety-five grams per day, or about 25 percent more than normal. The tendency to form kidney stones can be counteracted by liberal fluid intake and by alkalinizing the urine, for instance with potassium citrate. Although too much dietary calcium promotes calcium excretion and increases the risk of kidney stones, only one episode of stones has been reported in astronauts and cosmonauts through 2010.

Earlier, some NASA numbers were mentioned for the human body: 3.5 kg of water and 0.7 kg of food per person per day. NASA also allows thirty kg of water for hygiene, laundry, and other activities. Let's assume water for food and hygiene can be recycled and reused at 80 and 96 percent efficiencies, respectively. Also, neglecting differences in size, metabolic rate, personal preference, and the exact energy content of foods will not affect the outcome of the following exercise. In addition, daily body losses of CO_2, liquids (sweat, urine), and solids (salt, feces) are tallied and shown along with the requirements in figure 6.3. The difference between everything in and everything out is the *mass balance*.

Adding up the masses of O_2, H_2O, and food needed by an astronaut, the sum for one year in space is 0.9 kg O_2 plus 0.7 kg food per day times 365 days. This is 329 + 256, or 585 kg (1,287 lbs). Add 3.5 kg H_2O/day and 30 kg of hygiene H_2O/day, and after recycling at 80 and 96 percent respectively, allow losses of 0.7 + 1.2 kg/day. Thus, each astronaut needs 694 kg of surplus water for a year in space. The total mass per astronaut per year is therefore 585 + 694, or 1,279 kg (1.279 metric tons). For a three-year Mars mission, this value

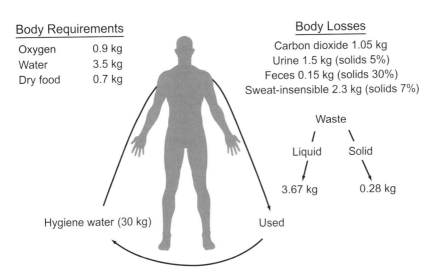

Body Requirements

Oxygen	0.9 kg
Water	3.5 kg
Dry food	0.7 kg

Body Losses

Carbon dioxide 1.05 kg
Urine 1.5 kg (solids 5%)
Feces 0.15 kg (solids 30%)
Sweat-insensible 2.3 kg (solids 7%)

Waste
Liquid   Solid
3.67 kg   0.28 kg

Hygiene water (30 kg)   Used

FIGURE 6.3. THE MASS BALANCE OF THE HUMAN BODY.
Daily requirements are based on an average metabolic rate for a 70 kg man in a thermally neutral environment (140 watts). Losses are apportioned into liquid and solid components, except for CO_2, which is a gas at body temperature. Hygiene water is for bathing and for washing clothing and other items. Solids can be recycled along with waste water.

is multiplied by three and then by six crew members, yielding some twenty-three metric tons. Now add up the three years of waste that must be stored or jettisoned: 85 kg of old food packages, plus 694 kg of brine and sludge per astronaut per year. This adds up to fourteen metric tons and does not include any reserves.

Finally, let's illustrate how much "junk" a crew on a voyage to Mars would generate if spent resources are recycled at an efficiency of 50 percent. On the positive side, the ship would carry only half the usual amount of supplies on the mission. In other words, a round trip requires exactly the same amount of O_2 as a one-way trip where there is no recycling. Of course, there's no such thing as a free lunch: more energy is needed, and the crew must still dispose of half of what it uses. The important exception is water, because of its more efficient recycling.

I once thought of waste disposal in space as just so much rubbish, only later recognizing how little trash is appreciated until the garbage man goes on strike. The cities of New York and Naples are notorious for this, but the *Mir* station was once euphemistically known as the "pigpen" in space. When resupply missions to the ISS were interrupted for two years after the *Columbia* mishap, garbage became a major issue. In 2004, Space.com even ran a headline, "There Is No Space on the Space Station."

It is not necessary to go into great detail on garbage disposal; a simple example will suffice. In 2006, a NASA committee on waste processing and resource recovery estimated that a crew of six en route to Mars would produce about 10.5 kilograms (23 lbs) of organic waste each day. A year would thus produce nearly four metric tons of biodegradable waste. The question is what to do with it.

The idea of returning tons of waste to Earth from Mars, some of it three years old, would be unpleasant for the crew, who would prefer simply to shoot it into space. On Mars, the expedient option, which would be to compact it and fill a crater up with it, is politically incorrect, but that is safer than trying to take off with eighteen months of trash. Trash dumps are not allowed at the South Pole, and they would be offensive on Mars too, but practicality may supersede sensibility. Technically advanced options like waste reconditioning for fertilizer or bioreactor fuel are expensive.

The idea of producing power biologically—so-called *biopower*—has been around a long time, but it is in its infancy for space exploration. Mature options include devices that convert kinetic energy (motion) into electrical current, like lamps powered by pedaling your bike or cell phones charged by cranking a handle. An interesting new prospect, microbial fuel cells, relies on bacteria that release electrons during metabolism. Both types of systems, as well as some that convert cosmic radiation into electricity, are on the drawing boards.

Microbial systems offer a handy way to store energy because of a high power-to-weight ratio. They are also a means of disposing of biological waste and a source of useful byproducts like O_2. The downside is that growing massive vats of microbes on an enclosed spacecraft or on another planet is hazardous and reckless, respectively. NASA thinkers estimate that the amount of biodegradable waste produced on a Mars mission could generate 1 kW of electrical power, and say a Mars ship's life-support system would require about 1 kW per person. Thus, a biofuel program might contribute 17 to 25 percent to life support and to waste disposal depending on the size of the crew. Microbial fuel cells also could be grown only when there is a critical mass of organic matter or when it reaches the point that the crew can no longer stand it.

The interest in space-based bioreactors has centered on bacteria of the genus *Geobacter* and species *sulfurreducens*, an anaerobic sedimentary microbe discovered in 1987 by D. R. Lovley of the University of Massachusetts. *Geobacter sulfurreducens* (as well as species found more recently) are called "electricigens" because they create electrical current by transferring electrons directly to an external electrode (Jones 2006). Electricigens produce energy by oxidizing organic material and transferring electrons to insoluble electron acceptors in their environments, like iron oxide particles in mud. Electron transfer occurs because the microbes produce conductive nanowires that ground it to the insoluble material—contact is made across the gap between the cell wall and the ground substance (Gorby et al. 2006).

In a working cell, trillions of bugs would oxidize waste in the absence of O_2 by using graphite as the electron acceptor instead of iron oxide. Since they do not require O_2, microbial fuel cells in principle would work on Mars. This is a tidy idea, but this too is still a field of research in its infancy, and we have almost no data on efficiency or durability. Energy is needed to support microbial growth, and it is not clear whether adequate trash disposal rates are attainable or sustainable at a reasonable scale here on Earth, much less in space. Given NASA's smaller budget today, there may not be time or money enough to develop this novel technology for use on a Mars mission.

7. ROUND AND ROUND IT GOES . . .
WHERE IT STOPS, NOBODY KNOWS

Biologists and engineers think of dynamic living systems as open *or* closed. In other words, the inputs are either consumed or conserved, regardless of anything else. However, every living system requires energy as a consumable input. An open system has access to all of its resources—O_2, H_2O, food—and consumes them. As resources are consumed, waste is excreted (or accumulates), and nothing is recycled apart from certain expensive biochemical factors and cofactors that do not concern us here.

Open systems operate in one direction, and waste is disposed of; we consume O_2 and hydrocarbons and eliminate CO_2, ammonia, some water, and heat. Metabolically, all higher animals are essentially open systems that use more resources than closed systems. However, the energy cost of impounding metabolic byproducts, managing these spent resources, and recycling critical resources is not trivial.

Since the early days of space exploration, when everything but the astronaut and his suit was disposable, sophisticated environmental control and life-support systems (ECLSS) have maintained a safe environment for astronauts. On the ISS, some recycling is done to conserve some resources. The ECLSS contains multiple subsystems that are integrated into a functional supersystem. Some parts of the supersystem are regenerative, and some are nonregenerative, but special control subsystems are required for both. The conservation of resources in space will demand increasing attention to detail in the future, tracking and replenishing the inevitable losses.

We just thought about water, O_2, and food, but CO_2 is critical, too, because it is needed to regenerate critical resources. It cannot be simply dumped overboard on long spaceflights. On the ISS, NASA has been adapting the Sabatier process for converting CO_2 and H_2 to H_2O and methane (CH_4). We will revisit this after examining the physiological context for why CO_2 must be conserved.

RECYCLING: OPEN OR CLOSED, HOT OR COLD?

To briefly reiterate, nonregenerative and regenerative life-support systems are independent of whether they have a nonbiological or biological basis. There are also hybrid systems. Nonregenerative systems are *open*, and regenerative and hybrid systems are *closed* or *partially closed* through the use of chemical and/or biological regeneration. For instance, O_2 production by water electrolysis is chemical, while photosynthesis is biochemical. If plants scrub CO_2 from and return O_2 to the atmosphere, the recycling is biological, but if CO_2 is broken down in canisters using a catalyst, the recycling is chemical. All such processes are ultimately chemical, of course, but it is a challenge to build manmade systems that mimic or improve on natural ones.

On the ISS, the ECLSS subsystems are monitored and tweaked, but in an ecological sense, even those capable of recycling are not self-sustaining. They are artificial because renewable biological components have not yet been incorporated into them. Regenerative terrestrial-like ecosystems for air, thermal comfort, water, and food for a crew on a long exploration mission may appear later in this century (if NASA hires back a few botanists).

Prototype biologically regenerative or *controlled ecological life-support systems* (or CELSS, in NASA speak) do exist, and several have been tried out, but their operation is challenging. There are no practical embodiments of these for long space missions, so I can only report the perspective of some experts on the requirements of future systems where resupply missions from Earth are impracticable.

The use of self-sustaining life-support technology far from Earth will require new knowledge of both natural and human-engineered processes.

This technology must set a well-defined barrier that prevents us from being exposed to conditions that are too close to a long-term survival limit. This means understanding the long-term biological effects of even modest deviations from our natural environment. For these reasons, life-support technology for deep space is not mature.

I have come full circle in recycling by revisiting thermodynamic efficiency. We are happy with machines, like automobile engines, that operate at around 50 percent efficiency. This is not bad; muscular exercise is only around 20 percent efficient. Still, there is promise in higher efficiencies, for instance, in fuel cells that "cleanly" convert hydrogen or hydrocarbons to electricity. A fuel cell's efficiency is defined by the highest efficiency that the reaction can theoretically achieve. Efficiency involves the entire system, all of its components, which means that each step is multiplied to arrive at an overall or *conversion efficiency*. The design of these systems varies, so it is not possible to generalize about efficiency. For instance, the theoretical thermodynamic efficiency of a hydrogen-oxygen fuel cell is 83 percent (Zhu and Kee 2006), and certain solid-oxide fuel cells (SOFC) and ultra–high efficiency solar cells have efficiencies up to 87 percent (Leya et al. 2005). In prototype systems, the conversion efficiencies are exceptional, 60 to 70 percent, but there are thorny practical issues to be worked out before implementation in space.

Living organisms are tiny next to their natural environments, so open biological systems are the norm. The entire lower atmosphere of the Earth, for instance, is available for you to breathe. Most animals operate as though the resources in their environments are limitless, since they are regenerated ecologically. Green plants in the photosynthesis cycle take up CO_2 and release O_2 into the atmosphere, where it is used by animals. So people are open, but Earth's biosphere, apart from sunlight, is closed.

Closed systems contain life resources for finite periods of time; the inputs are used, and the outputs are recycled as new inputs. This method decreases the dependence of the system on new resources, but not on energy. A purely closed system wastes nothing; it is *self-sustaining* or *self-renewing*, but external energy is always needed to support the cycles and the amount of energy needed exceeds that required to actually perform the work. The Earth's biosphere is sustained by the Sun, but it is not highly efficient—nor does it have to be, because so much solar energy is available (Pianka 1994).

Manmade systems have been open or hybrids of open and closed systems. For a particular subsystem, whether it is open or closed depends on the resource involved. For instance, on submarines, the reactor powers the electrolysis of seawater, which produces O_2 and H_2. H_2 is discarded, and the O_2 replenishes the atmosphere. Respiration reduces the O_2 to water, and the body excretes CO_2, which is dumped. This is an open system. If the CO_2 is chemically split into carbon and O_2, the O_2 can be reused and only simple carbon wasted. For potable water obtained by desalination, brine and wastewater is dumped overboard, where it is recycled by the ocean. Seawater is a limitless resource to submariners as long as the salt can be removed.

On the ISS, water is limited, and it is recycled, but H_2O is still used to generate O_2 by hydrolysis. The ISS backup systems are nonrenewable—O_2 cylinders and perchlorate candles must be supplied from Earth. In planning for Mars, more efficient regenerative systems are on the drawing boards, especially for the use of CO_2 and other waste products for food production.

Earth's ecosystems also fashion old into new and sometimes *better* or more successful things: in other words, they evolve. Our biosphere is not simply a huge recycling system; it is more precisely a set of linked ecosystems encompassing all life on the planet along with resources that sustain and change it. Apart from sunlight, the Earth contains everything necessary to support the life cycles of millions of diverse but connected species (Pianka 1994). This simple analysis makes it seem imperative to reconstruct a biosphere to sustain us in space. For more than a century, futurists have envisioned enormous closed ships that would carry us to the stars, and some even predicted that our Solar System has a carrying capacity of trillions of people (Daly and Townsend 1993). You can already see the flaws here, but let's bring on some more science on respiration, heat, and the CO_2 cycle aboard a spaceship.

THE CO_2 CYCLE

The principal source of human energy is *respiration*, and the lung is the interface that enables gases in the atmosphere to be exchanged with the blood and tissues. The heart pumps the blood through the lungs to the tissues, enabling the exchange of gases and nutrients. This simple description embodies the two components of respiration: external respiration, defined by the ventilation and pulmonary gas exchange, and internal respiration, which entails cell respiration and the delivery of O_2 and nutrients and the removal of metabolic byproducts, such as CO_2, from the tissues. Thus the lungs, heart, and blood link external and internal respiration.

Ventilation adds O_2 to the blood and removes the CO_2 generated by the breakdown of the hydrocarbons that supply the electrons that reduce O_2 to water. CO_2 diffuses from the cells into the blood, where the circulation carries it to the lungs, where it is exhaled. To live, we need constant access to a gaseous environment high in O_2 and low in CO_2.

Earlier, I mentioned that the cell's main energy supply is ATP, which contains three high-energy phosphate bonds that release chemical energy when they are broken. The most energy is released by breaking the first bond, which yields ADP and phosphate (P_i). ADP is recycled to ATP primarily by *mitochondria*, specialized organelles that utilize electrons derived from glucose and other substrates and add protons to reduce O_2 to H_2O. The process by which mitochondria convert ADP to ATP is *oxidative phosphorylation*, which produces most of the body's ATP. Mitochondria are fairly inefficient and therefore also generate heat.

In short, respiration irreversibly reduces O_2 to H_2O, ATP is produced, and heat is released — this is also why we are warm blooded. This heat generated by the body is measured by calorimeters much like those that measure the calorie counts on food packages in the grocery store. The oxidation of a gram of carbohydrate yields 4.18 kilocalories (kcal) of heat (or five kcal per liter of O_2). One gram of protein yields 4.32 kcal of heat (or 4.46 kcal per liter of O_2), and a gram of fat yields 9.46 kcal (or 4.69 kcal per liter of O_2). The quantities in parentheses are the basis for the so-called O_2 equivalent of power and the efficiency of muscular exercise.

The efficiency of exercise at constant work is defined by the ratio of how much heat is generated relative to the amount of O_2 consumed. When we exercise by walking, running, cycling, or swimming, our efficiency is around 20 percent (typically 16 to 24 percent for different exercises) (Jones 1997). In other words, about five times more O_2 is consumed than would be needed if all the energy from burning the fuel was used to perform the work; the extra is released as heat. There is thus a direct relationship between the amount of work done and the amount of O_2 consumed and amount of CO_2 produced per minute.

The average, supine, resting person consumes about three milliliters (ml) of O_2 per kilogram of body weight per minute. The body thus consumes 200 ml of O_2 and produces about 170 ml of CO_2 per minute. The resting O_2 consumption is often converted to a basal metabolic rate (BMR), or the number of calories per day needed to live. BMR varies with age, weight, muscle mass, gender, and genetic factors, and the Web is loaded with tables and calculators of BMR like the closely related BEE from chapter 6. Also, the ratio of CO_2 produced to O_2 consumed varies with the fuel used; this ratio is the respiratory quotient (RQ) and varies from 1.0 for carbohydrates to 0.7 for fats. For a person on a normal diet, RQ is usually about 0.85.

A seated person uses slightly more O_2—about 3.5 ml per min per kg of weight, because the postural muscles must perform work for you to stay in the chair. This value is called a *met*, and *mets* (multiple of resting metabolic rate) are used to describe the energy cost of various activities. For instance, sedentary jobs require two or three mets; occupations such as construction and mining require four to eight mets. Heavy exercise can use ten mets, and elite athletes involved in cycling, swimming, or cross-country skiing can use fifteen to twenty mets. Astronauts on EVA average three or four mets and reach peaks of six to eight mets.

During peak exercise, a highly trained, elite, male athlete can consume more than six liters of O_2 per minute. And depending on the fuel, this athlete will produce almost the same number of liters of CO_2 per minute. Such high loads are not sustainable for long, and in the exercise programs on the ISS, the work rates are usually less than six mets. To make a long story short, NASA allows 0.9 kg of O_2 per day per astronaut.

CO_2 is removed from the cabin atmosphere because of its undesirable effects. The CO_2 concentration in the cabin of the ISS, for instance, is in the range of 0.2 to 0.7 percent. If the inspired CO_2 increases to above 1 percent, breathing is stimulated, and the body's acid-base balance is altered. A chronic acid load (metabolic acidosis) may promote calcium excretion and worsen the effects of microgravity on the loss of bone mineral density (Bushinsky 2001).

The ability to remove CO_2 from the cabin quickly (and supply O_2) is critical to life support in a partially closed (or closed) atmosphere. The more work being performed and the greater the number of individuals, the greater must be the minute-by-minute capacity to supply O_2 and eliminate CO_2. The cabin is a *reservoir*, and it must be large enough to adequately buffer the peaks in the respiratory cycle of the crew.

One more thing about respiration and conservation: Recall that the respiratory cycle takes in O_2 and glucose and produces H_2O and CO_2. The H_2O contains the original atmospheric O_2 in reduced form. The O_2 in the CO_2 molecule is from the oxidation of glucose or another substrate. The cell converts hydrocarbon substrates like glucose to CO_2, passing the electrons off to O_2 to make H_2O, with an overall efficiency of about 20 percent. This O_2, although in two different chemical forms and from two different sources, is all available for recycling, but recovering it requires more energy than originally extracted as ATP and heat.

At the end of the book, we'll put recycling to a different litmus test—one based on extreme efficiencies without regard to technology. Dropping the technology and putting a process through its paces will find the limit on how far an approach can take us even with near-perfection—for instance, in trying to cross large, resource-poor regions of space. This has some eye-popping implications and puts starhopping into a rather unique perspective.

MAKING ROOM FOR THE JOLLY GREEN GIANT

Using CO_2 to grow plants to help solve food production and food-quality issues in space is an obvious strategy, but it is not as easy as it sounds. NASA food scientists have been working on this idea for a long time, particularly for a mission to Mars, where a significant portion of the crew's food (and possibly O_2) might be efficiently derived from plants.

When space food is prepackaged, the storage weight is much less if it is dehydrated and later reconstituted with recycled water. Such packages must have an extended shelf life. Thermally stabilized emergency rations on the ISS are good for at least two years, but this is not long enough for a Mars mission. Based on experience with submariners and powdered eggs, I agree with NASA food scientists: providing fresh foods, like vegetables and leafy greens, on exploration missions is a mighty good idea.

The issue of growing plants in space catches everyone's eye, and plant selection is based on lighting, photosynthetic capacity, size, variety, fertilizer, and edible yield (Smith et al. 2001, Drysdale et al. 2004). Experiments have been conducted on plants in space and in space simulators for years, and NASA has winnowed down hundreds of possibilities to a manageable number, including dwarf wheat, lettuce, tomatoes, rice, soybeans, and potatoes. These selections are based on high crop yields, food value, and hydroponic success—too bad for asparagus lovers.

Early studies on plant growth in space were focused on the effects of gravity on plant orientation, a phenomenon known as *geotropism*. In space, most plants germinate, develop, and grow normally, except that the roots and shoots do not orient themselves vertically. Some species show minor structural changes, such as leaf number and thickness, growth rate, or stalk height, related to microgravity, radiation, or both. The plants can be oriented using light.

Plants, at least for a few generations, also reproduce normally in space. The physical effects of microgravity are poorly understood, and research is needed to understand how space environments affect crop growth and reproduction. It is especially important to learn more about how cumulative time in space affects germination, growth, and development over many plant generations.

The effect of cosmic radiation on seed germination and seedling development is a crucial problem. A single, heavy, high-energy cosmic particle can wreck root development, for instance in tobacco plants grown from seeds, and break chromosomes in lettuce embryos (Planel 2004). Such problems are often silent until the seed germinates because the radiation causes mutations that affect plant growth or development.

A cosmic ray of high mass, such as the nucleus of a carbon or iron atom, passing through a cell, acts like a large-caliber rifle bullet passing through a watermelon. The destructive energy is extremely large. The tracks of such particles through cells were first documented aboard *Apollo 16* and *17* in the Biostack 1 and 2 experiments, which used encysted embryos of the *Artemia* shrimp of the Great Salt Lake. This tiny embryo can remain dormant for years but hatches into a nauplius when it is returned to high-salt conditions. The likelihood of failure to hatch and the nauplius to grow after a collision with a single cosmic ion in low Earth orbit is very high—up to 95 percent, versus a 50 percent hatch rate in Earth controls (Planel 2004).

The presence of photosynthetic food-producing systems would also support the O_2 and CO_2 cycles on long missions. The feasibility of this was first demonstrated in the 1960s in recycling studies conducted with algae and artificial light. It took an algae mat of only eight square meters to release enough O_2 to support one adult. Later studies have suggested that at least forty square meters of photosynthetic crops are needed per person to recycle O_2 and to provide food. Thus, a crew of six would require 240 square meters of photosynthetic plants, roughly one-half the area of a standard basketball court.

Adding a safety factor of 50 percent, we now have three-fourths of a basketball court. In a spacecraft, however, volume, not area, is the key factor. If the plants are stacked a half meter apart (including the hydroponic or dry medium), about 120 m^3 would be needed to grow them. The crew compartment of the *Apollo 13* capsule had a volume of only six cubic meters, and the *Orion* cabin is about twenty cubic meters. On a Mars mission, room must be made for plants, but lately NASA has been furloughing its botanists.

Since cabin space is likely to be tight, a two-tiered plan may be better for missions to Mars. The first tier is for the *transit* to and from the planet. Tier two is for the habitat. NASA scientists have envisioned a transit system similar to that of the ISS based on prepackaged cold or hot food trays with three-to-five-year shelf lives. This food system would provide a vegetarian diet similar to one available on Earth, but without eggs or dairy products.

On Mars, the crew would plant fast-growing crops in a special garden. In addition to those already mentioned, peanuts, legumes, spinach, herbs, and carrots could be grown. Fast growth in Martian greenhouses may actually be pretty slow, because Mars receives only half as much sunlight, more ultraviolet light (no ozone layer), and far more cosmic radiation than Earth. It may be necessary to provide dedicated microclimates enriched in CO_2 to accelerate plant growth and photosynthesis (Yamashita et al. 2006).

We know about Mars's soil composition from measurements made by the *Viking, Mars Pathfinder,* the Mars rovers, and the *Phoenix* lander. Like Earth, Mars soils are rich in metal oxides, particularly iron oxide, which imbues the planet with its famous redness. The levels of iron oxide, 13 to 17 percent, are more than twice those on Earth (Reider et al. 1997). Martian soils, however, contain little phosphorus and almost no nitrogen—tough going, even for lichens.

Most seeds would not sprout happily in Martian soil simply by adding sunlight, CO_2, fertilizer, and water. The ultraviolet, cold, aridity, and soil oxidants, including perchlorate salts discovered by the *Phoenix* lander in 2008, are not made for plant growth. Hydroponics might work well, but it is cumbersome, and if we choose it, we better have a good working surface technology to recover water on the Moon and Mars.

NASA crop scientists favor artificial soils developed from raw materials into which special nutrients can be incorporated. Such "space" soils resemble cat litter; they are porous and capable of absorbing and holding large amounts of water. Manmade soils are based on minerals like calcium apatite (hydroxyl apatite is the main mineral in bones and teeth) that contain or retain plant nutrients such as phosphorus (McGilloway and Weaver 2004).

Natural and artificial aluminum-silicate minerals, also called zeolites, are being widely tested, but today they are used mostly as molecular sieves (Mumpton 1999). Zeolites trap positively charged ions like potassium and ammonium, which can be leached out at a controlled rate. Under optimal conditions, artificial soils may actually outperform natural soils in terms of edible yield for some staples, such as wheat.

The growth of plants is only half the battle; sustainable technologies for harvesting and processing crops into palatable foods must be developed. Food production technology in space is inefficient with respect to energy utilization, but the barriers to achieving high efficiencies are not insurmountable (Drysdale et al. 2004). Time and research will solve these challenges, but crop science has a way to go before it can provide a reliable supply of fresh food on a Mars mission.

MICRONUTRIENTS

Micronutrients, especially vitamins, are another concern on long space missions. Micronutrients cannot be synthesized by the body and must be provided in the diet. Fortunately, they serve primarily as metabolic cofactors, and only tiny amounts of them are needed, hence the term "micronutrient." This aspect of nutrition in spaceflight is underappreciated because it is relatively easy to provide prepackaged vitamins and other micronutrients for short missions. For long exploration missions, micronutrients must be protected from oxidation and radiation for long periods of time by incorporating them into packets or pills. This means that the shelf life of these preparations must be extended significantly.

Of the thirteen vitamins, the four fat-soluble vitamins—A, D, E, and K— tend to accumulate in the body, while the nine water-soluble vitamins—the eight B vitamins (thiamine, riboflavin, folic acid, niacin, biotin, pantothenic acid, B6, and B12) and vitamin C—do not. Vitamin B deficiency diseases remain surprisingly common today, but the most interesting problem is with Vitamin D.

The vitamin D story is special because spacecraft cabins lack sunlight, and astronauts live in perpetual twilight (Rettberg et al. 1998). Vitamin D regulates calcium, phosphorus, and bone metabolism, and it promotes skeletal strength by facilitating calcium deposition in bone. Vitamin D deficiencies, especially rickets in children at northern latitudes, were once common because few foods apart from fresh fish contain it.

Even today, despite the routine fortification of milk and cheese, one in seven people have low vitamin D levels worldwide (Holick 2006, 2007).

According to the Harvard School of Public Health, people in North America who live north of the arc between San Francisco and Philadelphia and do not get fifteen minutes a day of sunshine are likely to have low vitamin D levels. Without vitamin D, only 10 to 15 percent of dietary calcium and about 60 percent of phosphorus is absorbed. Vitamin D deficiency in adults also causes bone diseases, such as osteopenia and osteoporosis, which increase the risk of fractures (Bischoff-Ferrari et al. 2004).

Vitamin D (D_2 or D_3) is normally acquired in two ways: it is consumed in the diet, and it is made in the skin. In the skin, a precursor compound called 7-dehydrocholesterol is converted to vitamin D_3, or cholecalciferol, by sunlight (UV-B radiation). Vitamin D from the diet or the skin is stored in fat cells and, when released, is metabolized in the liver to 25-hydroxyvitamin D, which the kidneys convert to an active compound called 1,25-dihydroxyvitamin D. This active vitamin binds to vitamin D receptors on cell membranes. The production of 1,25-dihydroxyvitamin D is regulated by the parathyroid hormone and by the plasma levels of calcium and phosphorus. Vitamin D also has important nonskeletal functions (Holick 2004). Skeletal muscles express receptors for vitamin D and may require vitamin D for optimal contraction. Vitamin D deficiency produces weakness. Vitamin D receptors are also present in the brain, prostate, breasts, large intestine, and immune cells.

Some two hundred genes are regulated by 1,25-dihydroxyvitamin D, including some involved in the regulation of cell growth, differentiation, and programmed cell death or apoptosis. Others are involved in new blood vessel formation, or angiogenesis. Vitamin D is also a potent modulator of immune function, and the immunological dangers of persistently low vitamin D levels are still being discovered. There are studies suggesting that low vitamin D levels are associated with an increased risk of developing prostate, breast, and colon cancer and with autoimmune conditions including multiple sclerosis.

The minimum RDA for vitamin D depends on age. The longstanding recommendation of 400 IU a day was increased to 600 IU for people ages one to seventy, but adults over seventy need 800 IU. In adults, the vitamin is well tolerated at intakes as high as 4,000 IU.

The optimal vitamin D intake in spaceflight, especially for a Mars mission, is unknown. Depending on how well vitamin D is absorbed, regular exposure to UV-B light may be needed too. Perhaps equally important are interactions between vitamin D and the development of radiation-induced cancers, but even for UV-induced skin cancer, there is essentially nothing known about this relationship. Moreover, other micronutrients, including trace metals like zinc and selenium, are also important for cancer prevention. Undoubtedly, micronutrients and radiation interact to influence the cancer risk, but at present virtually nothing else can be said about the problem.

Other important physiological effects of spaceflight are also probably linked to nutrition, such as decreases in muscle mass, red blood cell mass, and the antioxidant capacity of the blood. Changes in bone and muscle mass are attributable primarily to the effects of microgravity and, to a lesser extent, to changes in physical conditioning, similar to prolonged bed rest. The body's nutritional state interacts with these factors in ways that are unknown.

Vitamin K is found primarily in leafy greens but is also synthesized by bacteria that live in the intestinal tract. Although the body requires vitamin K for only one reaction, the so-called gamma carboxylation of glutamic acid, this function is critical to a number of proteins that bind calcium (Shearer 1995). This calcium binding is needed to initiate blood clotting, and the actions of certain anticoagulants, such as warfarin, antagonize vitamin K. This antagonism also decrease bone strength and predisposes to osteoporosis and fractures. It turns out that vitamin K is also required by the primary bone building cell, the *osteoblast*, which sets the skeletal strength not just on Earth but during exposures to microgravity and to partial gravity on the Moon and Mars.

In short, the proper intake of vitamins, trace metals, and other micronutrients in space is not simply a hypothetical issue. Diseases of vitamin deficiency have never disappeared on Earth, and we are likely to take them with us into space. Vitamin D is a special problem for life in twilight and for cancer prevention, and vitamins D and K interact to maintain bone strength.

8. BY FORCE OF GRAVITY

Every school child is told that Sir Isaac Newton discovered the law of gravity while sitting under an apple tree, but the grip of the ancients, dating to Aristotle, had been broken by Galileo's demonstration that heavier-than-air objects, regardless of weight, fall with a constant acceleration. This acceleration is caused by gravitational force, g, and the physical attraction between two objects. Gravity is proportional to the masses of the objects and to the inverse square of the distance between them. Objects dropped or thrown always fall to the ground, pulled toward the center of the Earth with a force of 1 g, or an acceleration of 9.8 m/sec^2.

Gravity, compared with the three other forces that hold matter together in the universe, is weak. It seems to dominate us because Earth's mass is enormous relative to the little bodies on its surface. Terrestrial objects have weight (mass multiplied by g), and although the law of gravity is universal, g is not. It is determined by the size and mass of the object. This means that the velocity needed to escape from an object increases with its size and mass. Values of g and for the escape velocity for the eight planets are shown in table 8.1. I call particular attention to Mars's escape velocity, which is 45 percent of Earth's.

ALLOMETRY

Gravity has a larger effect on biology than most people think. I mentioned geotropism in plants earlier, but gravity affects every living system by providing a frame of reference for directionality and orientation. It has also affected the evolution of everything from the paramecium to the largest terrestrial animals that ever lived. It is involved in all weight-bearing and load-driven processes, which has been recognized since antiquity. For all living organisms, the percentage of body mass devoted to structural support increases with the size of the body. This is the principle of *allometry*, first noted in 1538 by Galileo, who discerned that larger animals have proportionately thicker bones than smaller animals (Schmidt-Nielsen 1984).

The point is nicely made by the examination of the skeletons of the shrew and the elephant, the smallest and largest living land mammals. An elephant has four times as much bone in proportion to its mass as the shrew. Despite a bit of haggling over the arithmetic, about 14 percent of the elephant's mass is skeleton compared with only 4 percent for the shrew. We come in at roughly 8 percent, which coincidentally equates with marine mammals, from porpoise to blue whale, whose larger masses are offset by the buoyancy of water (Vogel 2003).

The rules of allometry set a limit on the absolute mass of terrestrial animals, which would collapse under their own weight if they were too heavy. The size limit for a terrestrial animal is related to the limits of bone compression, muscle strength, and to the stresses of locomotion. The limit for an animal based on bone strength, given an appropriate safety factor for getting up and walking about, is more than 10,000 (10^5) kg and probably less than 10^6 kg (Christiansen 2002).

TABLE 8.1. THE FORCE OF GRAVITY AND ESCAPE VELOCITY FOR THE EIGHT PLANETS

PLANET	RADIUS (EARTH = 1)	DENSITY (G/CM3)	GRAVITY (M/SEC2)	ESCAPE VELOCITY (KM/SEC)
Mercury	0.38	5.43	3.7 (0.38)	4.3 (0.38)
Venus	0.95	5.24	8.9 (0.91)	10.4 (0.93)
Earth	1.0	5.51	9.8 (1.0)	11.2 (1.0)
Mars	0.53	3.93	3.7 (0.38)	5.03 (0.45)
Jupiter	11.2	1.33	22.9 (2.33)	59.6 (5.3)
Saturn	9.4	0.69	9.1 (0.93)	35.5 (3.2)
Uranus	3.9	1.29	7.8 (0.79)	21.3 (1.9)
Neptune	4.0	1.64	11.0 (1.12)	23.5 (2.1)

This scaling effect is described in land animals by a simple geometric relationship, M^b, where M is the body mass and where the exponent b, the scale factor for the skeleton, is about 1.09. A scale factor of 1.0 means the skeleton grows in relation to volume. Values of b in excess of 1 indicate how much bone stress increases with size (Vogel 2003). This relationship nicely illustrates the effects of gravity. As for cetaceans, their value is about 1.02.

If you find exponents abstruse, listen to your local farmer or veterinarian, who will tell you that large animals are more likely to break a bone in a fall than small ones. "The bigger they are, the harder they fall" is true. The horse or the giraffe is far more likely to break a leg in a fall than the cat, which can alight unharmed from a fall from great height.

Elephants can weigh more than ten metric tons (10,000 kg), but this is still tenfold less than the theoretical terrestrial maximum. Many extinct animals of greater mass are found in the fossil record, and larger ones may yet be discovered, but the sauropods of the Cretaceous and Jurassic, some of which likely tipped the scales at 100,000 kg (220,000 lbs), may be near the limit for land animals.

On the Earth, gravity is masked by solid ground. The ground is solid because the electromagnetic and nuclear forces are much stronger than gravity; without these other forces, you would freefall toward the center of the Earth, 4,000 miles away. However, your acceleration would be undetectable without visual or other sensory cues. In fact, freefall with one's eyes closed simply causes a feeling of weightlessness.

Freefall accounts for the weightlessness of astronauts in low Earth orbit (LEO). Gravity is not absent; it pulls on the spacecraft holding it in orbit. However, for the astronaut, the force of gravity is cancelled out by the craft's forward acceleration, causing it to fall continuously along a path parallel to the curvature of the Earth. Thus, freefall creates an *apparent zero g environment*, even though the actual force of Earth's gravity, for instance on the ISS, is about 90 percent of normal. *Microgravity* is a state of freefall, an effect of approximately one-millionth of one *g*. To actually reduce gravity by this much (and not just simulate it in freefall), you must travel a distance of the square root of one million times the radius of the Earth (1,000 radii). Since the Earth's radius is 4,000 miles, this is about 4 million miles (6.4 million km), or roughly sixteen times farther away than the Moon.

Weightlessness in space was first predicted by Konstantin Tsiolkovsky, who sketched people and objects freely floating in spacecraft fifty years before the first spaceflight. Indeed, "weightlessness" is the earliest and most dramatic effect encountered by an astronaut entering orbit. The body reacts physiologically to microgravity, and the vestibular (balance), cardiovascular, and musculoskeletal systems are affected. Aerospace books devote entire chapters to how microgravity affects the senses, but since we are focused on *staying there*, these temporary troubles are of low concern. I will mention the vestibular effects briefly, but this chapter will primarily emphasize how microgravity affects bone and muscle on long-duration missions.

BALANCE AND PERCEPTION

Microgravity was originally predicted to have a major effect on balance and perception, given the spatial disorientation experienced by pilots of high-performance aircraft and airsickness in aircrews (NASA 2002). However, the inner ear (specifically the *vestibular labyrinth*) adapts surprisingly well to microgravity in a few days.

The labyrinth—the otolith organs and semicircular canals—operates cooperatively with eye movements and the position sense for the trunk and limbs. This gives us a sense known as proprioception. Proprioception provides the brain with information about the location and motion of the body in space.

The semicircular canals do not respond to gravity; they stabilize your vision when your head moves. The otolith organs sense gravity, body tilt, and acceleration through tiny stones called otoconia that exert gentle pressure on hair cells in the otolith organs—the utricle and saccule. These otolith organs are oriented along the X, Y, and Z axes of the head. The hair cells sense and translate gravitational information into electrical impulses, which the vestibular nerve transmits to the base of the brain.

On Earth, gravity provides a stable frame of reference and is normally sensed in the up-and-down direction. In aircraft, acceleration can come from any direction, and the nervous system cannot tell acceleration from gravity. In accordance with relativity, constant acceleration and gravity are equivalent. The sum of the two forces or vectors, direction and intensity, becomes a new "vertical" frame of reference.

The combined effects of acceleration and gravity create *vestibular illusions* that can cause aviation accidents, especially during a loss of orientation at night or in fog or clouds, where external visual cues are poor (Parmet and

Gillingham 2002). Pilots get a false perception, called a *somatogravic illusion*, of the *attitude* of the body with respect to the Earth. The most common is the *leans*, a false sense of displacement about the roll axis, or an illusion of banking, that causes the pilot to tilt in the misperceived vertical direction. Pilots are trained to ignore illusions and read their instruments, but the false sensations may persist for a while even after the directional forces have been corrected.

The lack of normal directional forces on the vestibular apparatus alters the perception of head position and head movement and leads to motion sickness. About two-thirds of astronauts develop mild motion sickness. The symptoms can often be minimized by avoiding rapid head movements, but individuals differ in their susceptibility, severity, and rate of resolution of motion sickness. There may be nausea, vomiting, and disorientation but rarely incapacitation, and an occasional astronaut may be unable to work for the first few days in space.

The brain adapts quickly to these gravitational changes, and the sense of equilibrium is restored as the illness subsides. Adaptation occurs because the brain learns to rely on cues from the eyes and the body's position instead of on the otoliths. Thus, much like seasickness, space sickness improves with time and responds to simple medications such as scopolamine or meclizine (Buckey 2006). Apart from dry mouth and drowsiness, these have few side effects.

Measurements of posture, gaze, and tilt after spaceflight show impaired gaze control, extra sway, and a reduced ability to control the tilt angle with eyes closed. These effects soon resolve, and chronic motion sickness is not seen in space or after returning to Earth. Several studies have indicated that the vestibular organs adapt to microgravity and may actually increase their sensitivity to gravity.

Motion sickness may also be encountered whenever gravitational forces change, for instance in going from microgravity to partial gravity after landing on the Moon or during the use of gravitational countermeasures to prevent the loss of bone and muscle. Chronic symptoms seem to be very rare, like the swaying *mal de debarquement*, which sometimes persists for days after a long sea voyage (Hain et al. 1999).

Space motion sickness is self-limited, and astronauts are given time to adapt before performing tasks taking concentration and coordination. Extravehicular activities (EVA) are not scheduled during the first three days of flight in order to minimize problems with disorientation while the astronaut is outside the spacecraft. Emergencies can occur any time, however, and the affected crew members may not respond optimally while they are still getting acclimatized to microgravity.

In the early days of the space program, apart from motion sickness, life scientists were worried about the effects of microgravity on the cardiovascular system, which proved more troublesome. Although they thought the heart and blood vessels would not tolerate weightlessness because the blood would lack weight, this fear was set to rest by early studies. However, other important cardiovascular effects of microgravity turned up, and some have long-term implications. Orthostatic intolerance was mentioned in chapter 2, and other important effects on the heart and blood vessels will be discussed later.

BONE AND MUSCLE

The loss of bone and muscle in microgravity has become a central issue in planning a trip to Mars (Turner 2000). These problems have been known for years, but the extent of bone loss and its progressiveness was not widely recognized until studies from the *Mir* station were reported in the 1990s (Vico 2000). The fact is that a strong skeleton is a lot more important on Earth than it is in space.

Since 2000, astronauts on the ISS have used vigorous exercise—lifting weights and running on a treadmill—as countermeasures for bone loss, but to little effect. This finding has been both puzzling and disappointing to life science experts. Measurements of bone density by computed tomography (CT) before and after visits to the ISS have shown that the rate of bone loss in the pelvis and lower spine is about 1 to 1.5 percent per month. Generally, the pelvis shows about twice as much bone loss as the spine (Lang et al. 2004).

Bone loss is such a concern because of the greatly increased risk of fractures after long missions. As people age, the skeleton thins and softens naturally, and the elderly are at an increased risk of osteoporosis. This disorder diminishes bone strength and is associated with a high incidence of fragility fractures. Osteoporosis has become more of an issue over the past century as the human lifespan has increased (Rosen 2003).

Osteoporosis is actually a disease spectrum in which several processes conspire to weaken the bones at different rates. The best understood factor is estrogen deficiency in postmenopausal women, which leads to deterioration of bone microarchitecture and to fragile bones. To talk about bone fragility, I must introduce the concept of bone remodeling, the major function of adult bone cells. Remodeling of the skeleton involves the removal of old bone (resorption) and the laying down of new bone (formation).

Bone resorption is the role of cells called *osteoclasts* and bone formation that of *osteoblasts*. Both types of bone cells precisely tune their activities to maintain bone health. Osteoporosis accompanies an increase in the rate of bone turnover or remodeling, which reduces bone strength. The weakening occurs because the phase of bone resorption is fast and that of bone replacement is slow. This produces a net loss of bone mass.

Since we naturally lose bone as we age, the loss of bone at a young age, for example in spaceflight, is associated with an increased risk of osteoporosis later in life. Major risk factors for osteoporosis, apart from spaceflight and estrogen deficiency, include immobilization, prolonged bed rest, low calcium intake, vitamin D deficiency or unresponsiveness, kidney disease, smoking, chronic inflammation, and the use of steroid medications (Raisz 2005).

The occurrence of fragility fractures is monitored by health organizations such as the World Health Organization and the National Osteoporosis Foundation (WHO Study Group 1994, NOF 1998). The higher the rate of bone loss, the higher the fracture risk, especially in elderly people. Most of these fractures occur from falls and commonly involve the hips and wrists.

If an astronaut returning from space suffers a large loss of bone, the rate of bone loss will return to normal, but the skeletal mass will not return to the preflight level. Thus, even after the rate has returned to normal, the bone remains thinner and the fracture rate remains higher. The loss of skeletal mass in microgravity has been refractory to preventive measures. Years ago, estrogen replacement was employed to prevent osteoporosis in women, but estrogen, particularly in combination with progesterone, increases the incidence of heart disease and breast cancer (Raisz 2005). Lower doses of estrogen also slow bone loss in postmenopausal women, but the net benefit over the long run has never been clear.

A newer class of drugs, the bisphosphonates, is widely prescribed now to prevent osteoporosis. These potent agents oppose bone resorption and are also effective in preventing bony satellite lesions or metastases in cancer patients. The drugs are taken up by osteoclasts, which are inactivated and die in an organized fashion called apoptosis, leaving the osteoblasts intact. Bisphosphonates do reduce the frequency of fragility fractures, especially hip fractures (Raisz 2005), but they may inhibit bone remodeling excessively and interfere with the repair of microfractures, leading to another form of bone fragility called *osteopetrosis*, which is usually a congenital disease. This makes the drugs more dangerous for younger people, and other side effects, especially heartburn and esophageal irritation, make them undesirable in microgravity. However, the bone loss in space is so severe that these drugs are being studied on the ISS to determine if they are helpful.

Astronauts also lose muscle mass during spaceflight, and this wasting or atrophy begins in just a few days. Without exercise, the wasting continues until the astronaut loses 30 to 50 percent of her muscle mass (Fitts et al. 2000). The affected muscle groups are primarily the postural or antigravity muscles, which are involved in standing, walking, and lifting, especially in limb extension rather than flexion. Astronauts on long missions may reduce this skeletal muscle loss by half by keeping up a vigorous exercise program.

At first blush, you may wonder if muscle atrophy in space is simply attributable to a gradual loss of muscle size and strength because the muscles are "unloaded," as with the immobilization of a fractured limb or with prolonged bed rest. This idea is actually sound, and aggressive exercise countermeasures in space have led to significant improvements (Greenleaf 2004). Crews on the ISS participate in a daily two-hour exercise program six days per week, although most still develop significant muscle loss. Thus, deconditioning is not the whole story, and to see why requires a closer examination of muscle physiology.

The skeletal muscles make up the largest tissue in the body, some 40 percent of body mass. Muscles are bundles or fascicles of *fibers*—long cylindrical cells—arranged in parallel along the long axis of the muscle. Each fiber is composed of a hundred or more smaller units called *myofibrils*. These contain protein filaments of actin and myosin, which slide by each other when the muscle contracts.

Muscle contraction is initiated by activating the motor nerve to the muscle, which stimulates channels on muscle cell membranes to allow sodium to enter the cell and trigger an electrical action potential. The action potential initiates a contraction. The contraction is coordinated by calcium ions released from storage sites inside the muscle cell. Calcium binds to a protein called troponin C on the actin filaments, and troponin C changes the shape of the protein filaments, causing them to shorten (Guyton and Hall 2000). Muscle contraction and relaxation also require energy, supplied by ATP.

There are two main muscle fiber types, white or fast fibers (type II), which primarily use glucose, and red or slow fibers (type I), which contain many mitochondria that consume O_2 to make ATP. The amount of O_2 delivered to the muscle is proportional to blood hemoglobin content and to blood flow, and the delivery of O_2 is facilitated by a red protein, myoglobin, closely related to hemoglobin. Myoglobin helps O_2 diffuse from capillaries into muscle cells and mitochondria. Of the two fiber types, fast fibers contract more rapidly and with greater force, but they fatigue more easily than slow fibers. Slow fibers develop force more gradually but maintain it for longer periods of time. As a rule, strength training conditions fast fibers; aerobic training, such as running and swimming, trains slow fibers.

The force of muscle contraction is converted into mechanical work. Force acting on a joint creates torque that rotates the joint on its axis, such as flexing the knee. During each contraction, one of three things can happen to the length of the muscle. If it shortens as a load is moved, the contraction is *concentric* or *isotonic* (constant tension). If muscle position is fixed, the muscle remains at a constant length, and the contraction is *isometric*, as in straining. If the muscle lengthens or stretches, the contraction is *eccentric*. All three types of contraction are important to training and conditioning, but microgravity, in addition to unloading the postural muscles directly, causes the loss of stretch or eccentric contraction. The loss of stretch is attributable to the loss of postural load, and it is an important factor in muscle atrophy (Convertino 1991).

Muscle stretching helps counteract the atrophy of spaceflight, and keeping stretch on the antigravity muscles in space is highly desirable. The Russians even invented an elastic suit called the penguin suit to do just that. It was tested in space, primarily on the *Mir*, but most astronauts found it uncomfortable, and its effectiveness is questionable.

As a boy, I was amazed that kids my age but stronger than me tended to be the lankier ones. It was quite a mystery until my father told me they had leverage. I didn't quite believe him until later, when I learned something more about strength. The strength of a contraction, or peak force, is related not just to muscle mass but to the length of the muscle, regardless of the degree of shortening. Strength is usually measured by a single maximum contraction, but power is the rate of performance of work or the amount of work the muscle does over time (per second). In other words, strength is peak muscle power, and endurance is power over distance. On long spaceflights, endurance exercise does not prevent atrophy. All it does is use up more O_2 and put more CO_2 through the life-support system.

Strength is related to muscle mass because large muscles have more actin and myosin filaments for recruitment during a contraction. The size of these filaments, which is related to the fiber cross-sectional area, generally determines the strength. Training increases both the size and the number of muscle fibers, but it also transforms one type of fiber into another; for instance, weight lifting increases the number of type II muscle fibers by allowing new fiber development and by converting red fibers into white fibers. This is called fiber-type switching, or plasticity. Endurance training will develop more red fibers and increase the density of the capillaries, the myoglobin content, and the number of mitochondria.

When muscle activity is interrupted, for instance after an injury, immobilization, or after a period of physical training, muscle performance gradually declines. Over days to weeks, immobilization and deconditioning reduce both strength and endurance. The largest and most rapid decrements occur after an injury with muscle immobilization, followed by immobilization of healthy muscle, and then prolonged bed rest, which is similar to spaceflight. The worst case involves muscle paralysis, which has to do with both disuse and loss of neurotropic (nerve) factors.

The degree of muscle atrophy is related to the extent and duration of disuse, and all fiber types show similar decreases in cross-sectional area. When the effects of injury, immobilization, and deconditioning are compared, there are a few biochemical differences in muscle atrophy, and this is also true of muscle atrophy in microgravity.

After an immobilizing injury, muscle strength usually decreases by about 50 percent but then stabilizes after about a month. If a limb without muscle injury is immobilized, performance declines more slowly, but the final loss of function is still about half. With simple deconditioning, muscle performance is stable for about a week and then begins to decline gradually. Studies of deconditioning have also indicated that the loss of adaptation is local and not attributable to neurotropic factors. In other words, if a muscle is immobilized or its training stops, the fall in muscle strength is caused by intrinsic changes in the muscle cell.

Prolonged bed rest also has an effect beyond that of simple immobilization, and it raises an important concern about microgravity. After three weeks of bed rest, the maximal O_2 uptake during exercise, the VO_{2max}, may decline by a third. Much of this effect is related to a decline in the cardiac output during exercise, attributable primarily to a decrease in the heart's *stroke volume*.

Stroke volume is the amount of blood put out by the ventricles during one heartbeat, usually about 1 ml per kg of body weight. The reason stroke volume decreases in space is complex, but the atrophy of heart muscle itself plays a very minor role. Instead, stroke volume falls because less venous blood returns to the heart and because the ventricles become stiffer or less compliant. These factors impair the ability of the heart to fill with blood during relaxation, or diastole.

The fall in the heart's stroke volume in space is comparable to that associated with protracted bed rest. The physiology, however, is more complex because of the low blood volume in space. We saw earlier that astronauts returning from microgravity are prone to fainting because the amount of venous blood returning to the heart is low. The stroke volume is low but can be improved by giving extra fluids to the astronaut before reentry. In microgravity, the blood was displaced centrally into the chest, and the heart's stretch receptors were activated, causing the body to undergo diuresis, similar to immersion in water. This response kept the volume of blood returning to the heart and the stroke volume near normal, but only while in space.

Definitive measurements of cardiovascular performance in space are not available, although the mass of the heart seems to decrease by only about 10 percent (Perhonen et al. 2001). This means that atrophy of the heart is distinct from atrophy of the skeletal muscles and that microgravity has a bigger effect on skeletal muscles. More studies are needed to understand how these physiological mechanisms operate, particularly on long missions.

The loss of muscle mass in spaceflight can be tracked indirectly by monitoring changes in body weight and muscle circumference or muscle volume, but the precise effects of microgravity on performance are not known because of the limitations of conducting research in spacecraft environments. Muscle volume measurements by magnetic resonance imaging (MRI) indicate that the volume of the antigravity muscles declines by about 20 percent, for instance, in the soleus muscle in the leg. For an individual, changes in muscle volume are a reasonable index of changes in performance, including strength and endurance.

On the ISS, the crew has a treadmill, a bicycle ergometer, and resistance trainers. Exercise regimens vary, and moderate aerobic exercise is performed for about five hours a week and resistance exercise on three days. Despite this regimen, after six months, astronauts show a significant loss of calf muscle mass and a one-third decrease in peak calf power along with the conversion of slow to fast fiber types in the gastrocnemius and soleus muscles. These are features of unloaded muscles (Trappe et al. 2009).

On long missions, muscle changes may be confounded by diet, by low vitamin D levels, and by changes in the levels of steroid hormones such as testosterone and cortisol. These effects have not all been sorted out, but postural muscle atrophy is greater than is seen with simple bed rest. The muscles of the upper extremity are affected less than those of the lower back, thighs, and legs because arm use increases and leg use decreases in microgravity. The use of the arms is the easiest way for astronauts to move around the spacecraft.

If the muscle atrophy of spaceflight involves cellular changes, muscle cells should look and behave differently. Simple early experiments made scientists think microgravity had little or no effect on muscle cells, or on any cells for that matter, because the forces that govern how cellular macromolecules interact are so much stronger than the force of gravity. Most of the organelles in a cell are slightly denser than the cytoplasm but too small to be affected seriously by microgravity. The nucleus and the mitochondria may be exceptions, but because they tend to be tethered to the cytoskeleton, the effects of microgravity on them were thought to be minor. However, more sensitive later studies have given some cause for concern.

Muscle atrophy and its doppelganger, hypertrophy, reflect imbalances between the breakdown (*proteolysis*) of existing muscle proteins and the synthesis of new ones. For instance, weight lifting increases muscle mass by increasing the synthesis of new muscle protein compared with proteolysis. In contrast, protein synthesis decreases and proteolysis increases during muscle atrophy. Extra nitrogen is released as proteins break down and is excreted in the urine, and this has been observed in microgravity environments.

In the absence of disease, muscle atrophy ultimately stabilizes, the synthesis and breakdown rates of proteins fall below the normal level, and the balance is restored. The contractile proteins are lost at a greater rate than other cell proteins, and the thin actin filaments decline more than the thick myosin filaments. This is not too surprising since less protein synthesis and more proteolysis are also hallmarks of malnutrition, neuromuscular wasting, hibernation, and acute and chronic infections.

Two other facts about muscle atrophy are often overlooked. The first is that disuse is associated with smaller and not fewer muscle fibers, indicating that idle muscle cells do not die. The muscle cross-sectional area and fiber density return to normal on reconditioning. The second is that type I or slow fibers, which are rich in mitochondria, are more sensitive to atrophy than type II or fast fibers. In fact, in some muscles 90 percent of the atrophy occurs in the slow fibers (Kalpana 1998).

During atrophy, hybrid fibers appear, and they revert during reconditioning. Type I fiber atrophy is also accompanied by a loss of muscle capillaries and mitochondrial density. Most antigravity muscles are loaded with slow fibers, which might seem odd—except that they normally contract and relax sixteen hours a day. Physiologists would like to know whether microgravity affects the formation of new muscle mitochondria and capillaries and, if so, how.

One way to learn more about how microgravity affects skeletal muscle cells is to study how spaceflight affects the expression of the genes that regulate muscle proteins. This means conducting work on rodents because muscle biopsies are easier to obtain from space-flown rats and mice than from astronauts. Let's look at the kinds of muscle genes that are turned on and off in microgravity. This can be done using the molecular profiling tools of *microarray analysis* and *genome wide association studies* (GWAS). GWAS allows for a simultaneous analysis of the expression of the thousands of gene products or messenger RNA molecules that encode for different proteins in healthy versus injured or diseased tissues.

GWAS is quite powerful, and it is helping shape modern molecular medicine. Certain diseases, such as cancer, diabetes, Alzheimer's, and cardiovascular disease, are being defined by the customized profiling of the genes expressed or repressed by cells of individuals at risk or already affected by disease. Gene profiling has also been applied to the study of skeletal muscles and cells in spaceflight in order to see microgravity effects (Nichols 2006).

One study in 2003 showed a decrease in the expression of metabolic enzymes involved in how mitochondria oxidize fats (fatty acids) along with an increase in the enzymes of glycolysis (Stein and Wade 2003). Differences have also been found between microgravity and simulations of microgravity, for instance, by gently suspending a rat by its tail so gravity does not act on the hind limbs. Another study in 2004 showed that microgravity affected the expression of 257 genes (100 went up and 157 went down), whereas tail suspension only affected seventy-four genes (thirty-seven up and thirty-seven down). Thus, spaceflight up- or down-regulated about 2.7 and 4.2 times as many genes, respectively, as did tail suspension (Nikawa et al. 2004). These two stresses are therefore not as comparable as once believed.

Microarray data from space-flown rats also shows two distinct gene patterns. The first is the unbalanced expression of genes that duplicate mitochondria along with the disturbed expression of proteins of the cell's internal scaffolding or *cytoskeleton*. These cytoskeletal proteins, such as dynein, anchor mitochondria in place. Dynein expression is disrupted by microgravity, and this changes the microscopic internal skeleton that governs the distribution of mitochondria.

The second pattern is the overexpression of the enzymes of protein breakdown (proteolysis). These genes activate the cell's main protein disposal pathway through a protein called ubiquitin and a pathway called ubiquitin-dependent proteolysis. Ubiquitin acts as a flag, attaching to damaged or abnormal proteins and identifying them for degradation. Flagged proteins are broken down into amino acids in organelles called *proteasomes*. The proteasome excretes or recycles amino acids for the synthesis of new proteins.

In muscle cells, mitochondria can be stained with vital dyes to assess their numbers and distribution. The staining patterns in the muscles of space rats indicate that exposure to microgravity leads to an abnormal distribution of mitochondria in the cell. Thus, microgravity seems to disrupt the spatial organization of muscle cells by interfering with the expression of proteins that anchor mitochondria to the cell's internal skeleton.

The distribution of muscle mitochondria may also reflect a gravity-sensing network that contributes to the rapid and characteristic structural and functional changes that mitochondria undergo in response to certain environmental stimuli. As this area of research expands, it may demonstrate intrinsic physical effects of microgravity on muscle structure and function that are not amenable to routine countermeasures. This may mean it will be necessary to restore part or all of the normal force of gravity on long space missions.

WEIGHTLESSNESS AND THE WHITE BLOOD CELL

Forty years ago, one of the great puzzles of spaceflight became clear: microgravity affects immune function in an odd way. Spaceflight interferes with the ability to activate and populate certain types of white blood cells (Sonnenfeld and Shearer 2002). In the 1960s, Apollo astronauts were found to have mildly depressed immune cell function after their flights, and this took about a week to recover. It was noticed that the circulating T lymphocyte could not be activated. T lymphocytes are central to cell-mediated immunity and comprise an important first-line defense against microbial pathogens.

At the time, it was thought that this effect was caused by changes in stress hormone levels or perhaps by inadequate nutrition; however, in 1984 data were published demonstrating that T cells also could not be activated properly in space, implicating microgravity itself as the cause of the problem. Subsequently, it was observed that specific key immune-response genes in lymphocytes do not respond to an activating stimulus in microgravity. Why this loss of gene regulation occurs is unknown.

Practically speaking, astronauts are not overly impaired by weakened immune responses. They do get colds and other minor viral illnesses more often than normal after returning from spaceflight, but as people spend longer periods of time confined to spacecraft or habitats, there may be an increased risk of activating latent infections or contracting new infections. This aspect of space biology requires more understanding.

The complexity of the immune system is such that some of the effects of microgravity on it might not be primary but secondary or derivative. To explore this possibility, NASA funded some investigators to fly fruit flies on the shuttle and measure the behavior of their relatively simple immune systems. The fruit flies were housed and bred on the shuttle for thirteen days, and significant depression of their immune function was found, but mainly in larvae born in space (Marcu et al. 2011). This is further evidence that microgravity is the culprit.

This is not the only possibility. There is evidence that spaceflight increases oxidative stress both in the immune system and in the musculoskeletal system. Oxidative stress can be triggered in many ways, but on a spacecraft, two factors, radiation and microgravity, have been considered the most important. Oxidative stress increases the net production of reactive oxygen species (ROS), which damage the cell by chemically altering the functions of macromolecules including lipids, proteins, and nucleic acids. Interestingly, some of these events involve neuroimmune modulation (NIM).

NIM is a part of the body's responses to environmental stress, including those of isolation and confinement. Virtually every major stimulus to the central nervous system has secondary effects on both arms of the immune system. This can increase the susceptibility to infection, to autoimmune diseases, or to cancer. These are good reasons for understanding the effects of physical stimuli on NIM (Friedman and Lawrence 2002), including changes in the gravitational and magnetic and fields around the body.

Evidence is mounting for strong bidirectional communication between the brain and the immune system implemented by a range of hormones and small proteins, or neuropeptides, involved in the hypothalamic-pituitary axis (HPA) (Kim and Sanders 2006). Many stresses that interact with each other also influence the extent, and in some cases, the direction of NIM. Such interactions include the genetic background of the individual, the level of chronic stress, and the intensity, duration, and types of acute stress.

Immune function is linked to life's early events, and we know that this relationship continues throughout life. This immunological memory is driven by what are probably both preexisting genetic programs and acquired environmental or epigenetic programs (Moynihan and Santiago 2007). There is the disturbing possibility that a life spent in microgravity or partial gravity might significantly disrupt the regulation of our endocrine systems. Some investigators have suggested that microgravity interacts with light levels, for instance in winter twilight, but this has not been studied carefully. Snippets of information do provide a provocative glimpse of how the body's endocrine performance and biological cycles change in extreme environments.

Chemical compounds called endocrine disruptors, which are being found everywhere today, are generating public health concerns. Some occur naturally, but many are manmade, such as bisphenol A, an ingredient of plastic implicated in cancer (Jenkins et al. 2011). These disruptors mimic natural steroid hormones and interfere with normal steroid hormone synthesis and function. Some affect reproduction, others the functions of steroid-sensitive tissues like the thyroid and the brain. There are chemicals that affect human immune function, some that cause diabetes, and some with epigenetic effects that modify gene expression and have transgenerational effects. The implications for closed environments are enormous, but untested.

The human body maintains a vestige of seasonal thyroid hormone secretion related to intermittent exposure to the cold. This can be detected, for instance, during residence in Antarctica (Do et al. 2004, Burger 2004). Long-time residents of Antarctica acquire the "Antarctic stare"—episodes of gazing impassively off into space for no apparent reason. The stare is associated with mild changes in thyroid function known as the polar T (3) syndrome. The syndrome is characterized by an increase in thyroid gland function shown by increases in circulating levels of thyroid stimulating hormone (TSH), the hormone-binding protein thyroglobulin, and changes in the production and clearance of active thyroid hormone—free thyroxin (T4), which falls, and T3, which rises.

The cause of these thyroid effects is not entirely clear, but the cold climate is an obvious factor, particularly the increased utilization of thyroid hormone by peripheral tissues in the cold, especially in skeletal muscle, as well as less thyroid hormone binding to thyroid hormone receptors on the cells in the brain. The loss of thyroid hormone effects on the brain adversely affects the levels of neurochemical transmitters and could lead to depression of both immune and cognitive function.

There is mild clinical depression and cognitive decline observed during long stays in Antarctica, which thyroid supplements appear to alleviate. This suggests that changes in thyroid hormone synthesis and interactions in unusual environments are physiologically important and may operate through NIM. The implications for long-term effects on endocrine and immune function in spaceflight are again enormous but untested.

Our understanding of the physiological aspects of immune function in space is thus rudimentary, and more knowledge is needed about the regulation of the body's inflammatory and anti-inflammatory responses in space. This includes the effects of endocrine and nutritional factors and oxidative stress on immune function (Tracey 2005) and their interactions, in particular, with microgravity and radiation.

Physiological and biochemical equipoise in the immune system have important implications for resistance to infection, brain function, bone and muscle health, cardiovascular disease, cancer, and reproduction both on Earth and in space. The independent effects of microgravity on animal reproduction are, of course, also a source of both great interest and heated debate; the latter reflects a paucity of clear studies on animal gestation in microgravity, which to date primarily involve the zebrafish. In short, a better understanding of the long-term health effects of hormonal factors, oxidative stress, and chemical disruptors on immune function in microgravity and partial gravity is necessary because they may specifically influence the success or failure of prolonged space missions.

9. THE COSMIC RAY DILEMMA

So far, the challenges of human space exploration may seem to you as if they could be managed through technological and biological innovation. This is a reasonable supposition if we are tenacious enough and if costs can be managed, and the opportunity for future space exploration and exciting new spinoff technologies seems high. However, we have yet to discuss the silent but sizzling showstopper, *ionizing cosmic radiation.* Unlike the earlier problems, protection from cosmic radiation is a matter for neither simple hard shells nor biological adaptation.

This cosmic ray dilemma, as I say, will take time to solve. Some top radiation physicists think it is impossible, because the shielding requirements are so high, to eliminate significant ionizing radiation exposures in space and that the only practical strategy is to keep radiation exposures *as low as reasonably achievable.* This catchphrase goes by the NASA acronym ALARA. We will examine the dilemma in sufficient detail to understand it without getting stuck in atomic physics, but I want to be clear at the outset that our options for dealing with ionizing radiation are limited.

In 2011, just before Thanksgiving, a visitor from the Johnson Space Center spent the day hobnobbing with us in our laboratory at Duke University.

NASA visitors always have a unique perspective, and not being an inside man, I often pick up fascinating tidbits about how the agency operates. For years, I've detected a shared frustration with NASA's ponderous inefficiency as well as with aerospace industry diktats over space hardware.

Our visitor, Jim Logan, was no exception. A no-nonsense physician who has spent twenty years at NASA, Logan is an expert in telemedicine who even took a midservice hiatus in the private sector to spinoff some of his creative ideas. On the day of his visit, after a brief update on the use of remote heart ultrasound on the ISS, our conversation drifted toward the future of human spaceflight. He was less sanguine about spaceflight outside the Van Allen belts than I was mainly because he thought NASA would continue to waltz around the radiation issue for years to come. "It's like space hardware development and the aerospace industry," he said. "They grumble about how little we know, get a lot of money to study it, then come back for more money with a new list of problems."

"This is understandable to a point," I said, "but in time, shielding technology will settle the issue in deep space, and remote mining technology will solve it on the Moon."

"Yes," he replied, "this can be done. You saw how submariners solved the radiation protection problem, and they did it before anyone ever went into space. And I've always said that for Moon-Mars, we will have to start out like ants, earthworms, and moles—living underground."

Jim Logan is exactly right: Admiral Rickover's nuclear engineers solved the undersea reactor radiation problem in the 1950s. Twenty-five years later, I was amazed to learn that submariners on patrol had been living and working within one hundred feet of an atomic reactor and receiving less radiation than they did from natural sunlight during shore duty. However, as you will see, cosmic radiation is a unique problem.

Logan also emphasized the importance of nonpotable water in space. "In principle, water shielding combined with naval reactor-propulsion technology would work in space, with water both as shielding and as a propellant," he noted. Steam power may harken back to Jules Verne, and water is heavier than and not as "up to the minute" as solar-electric propulsion (SEP), but there is no reason that this well-established technology couldn't be adapted for spaceflight.

The problem of inefficient technology development is old news to NASA people, who nevertheless seem resigned to it. The 2011 Human Exploration Framework Team (HEFT) had noted that to maintain "affordability and shorten the development cycle, NASA must change its traditional approach to human space systems acquisition and development." They emphasized a flexible "capability-driven approach" to human space exploration, rather than one based on specific destinations and schedules, and they stated that a fifteen-year exploration horizon was too short.

Reading between the lines, the Achilles' heel is clear: the lead time for new technology is too long, and the mission timeframe is too short; hence, long delays, cost overruns, and frustration. However, the advice of the HEFT is eerily similar to that of the two dozen other extramural panels appointed to advise NASA since Project Apollo. It is not clear how to make the whole thing match up in today's political and economic climate, but there is a fresh expectation that the move toward privatization will increase both the competitiveness in the aerospace industry and competitively priced options for NASA.

I have wandered afield now, since our topic is cosmic radiation, but the radiation problem is a logistical nightmare, which affects our ability to solve it. Earlier I mentioned three natural mechanisms that have done a remarkable job of shielding life from cosmic radiation for billions of years: Earth's atmosphere, oceans, and magnetic field. We can exploit these same mechanisms in space.

The types and intensity of radiation in the sky vary with latitude, altitude, and the phase of the solar cycle (Townsend 2001). The details create uncertainties in the actual dose of damaging cosmic radiation, but these details aside, the importance of the atmosphere in radiation protection is easy to demonstrate in round numbers. If we arbitrarily set the natural background radiation at one unit of biological effect per year, the dose, because of the solar and galactic radiation components, gradually increases with altitude. Atop Mt. Everest (29,029 feet, or 8,848 meters), we are receiving about one hundred units. Although no one stays on Everest very long, commercial and military flight crews who fly seven hundred hours a year receive at least one hundred times more cosmic radiation than people who live in New York City. The airlines have noticed because the risk of cancer in these flight crews has been estimated at close to 1 percent above the normal lifetime risk of 25 percent.

There is a hidden factor in this statistic too: the level of O_2 inside commercial aircraft. This level is maintained at the equivalent of being at a slightly higher altitude than Colorado Springs. In theory, O_2 can raise (or, in hypoxia, lower) the radiation cancer risk by interacting with radiation to influence the genetic stability of the cell through increased reactive oxygen species (ROS) and other free radical mechanisms. The interaction of O_2 with radiation is defined by the *oxygen enhancement ratio*, which indicates that radiation-induced DNA instability is increased by about threefold by O_2. Regrettably, there is no hard information on the concentrations of O_2 required to affect the radiation cancer risk in people in hypoxic environments.

If you recall the genetic mutation the Tibetans acquired to help them adapt to high altitude, you know that 90 percent of Tibetans but only 10 percent of the Han Chinese carried it. My colleague "Q," who happens to be of Han descent, has his eye on joining the crew of a Tibetan starship operating at two-thirds sea level O_2 and fifty times more radiation than astronauts experience in LEO. This seems like ALARA in the extreme, but his Darwinian reasoning is based on lowering the O_2 enhancement ratio. By having seen more radiation and less O_2, his descendants might eventually become better adapted to radiation. Small epidemiological studies suggest "Q" might be right (Amsel et al. 1982), but his experiment of nature won't be performed any time soon—and we'd also have to wait centuries before we could look for the results.

THE ABCS OF COSMIC RAYS

The effects of ionizing radiation are linked to a fairly simple physical chemical principle: radiation interacts with the molecules in living systems and modifies them chemically so that they don't function properly. This molecular radiation damage is often permanent. A radiation-damaged molecule must be repaired or replaced in order for the affected cells and organs to regain the function and to avoid adverse consequences later on. These consequences are known simply as late or delayed radiation effects. Radiation, like ROS, damages all sorts of biomolecules, including proteins, lipids, and nucleic acids.

The consequences of radiation exposure run the gamut from acute radiation sickness to later-appearing effects like cataracts, brain degeneration, vascular disease, birth defects, and cancer. In deep space, without our protective atmospheric umbrella and Van Allen belts, the lives of astronauts are endangered by both the acute and the late effects of radiation.

As a general rule, higher radiation doses mean higher rates of cellular damage. There may also be a threshold dose below which significant biological effects either do not occur or are repaired completely. Whether such a threshold is important depends both on how the damage is assessed and the type of radiation, and there are many types. Space radiation is also qualitatively different from the natural background and medical radiation encountered on Earth. To paraphrase Orwell's Napoleon in *Animal Farm*, all forms of radiation are equal, but some are more equal than others.

Space radiation has not wiped out life on our planet because our atmosphere, oceans, and Van Allen belts absorb and deflect a spate of cosmic rays. Cosmic radiation is far more intense than background radiation and includes cosmic particles, a few of which do penetrate our atmosphere, especially at higher altitudes and at the poles. Principally however, we are exposed to that type of electromagnetic radiation known simply as *light*.

The wave property of light and the spectrum of electromagnetic frequencies are used to divide radiation into different types (see figure 9.1). The higher the frequency of the wave (shorter wavelength), the more energy it transmits. For instance, gamma rays, having a wavelength of a billionth of a centimeter or less, have the highest energies and can produce widespread molecular damage. Space contains the full spectrum of electromagnetic radiation, but the group of highly energetic particles, ions, and atomic nuclei called *ionizing particle radiation* is especially hazardous to life.

Recall that atomic nuclei are made up of two types of baryons: positively charged protons and uncharged neutrons of almost exactly the same mass. These are held together to form the nucleus by the strong nuclear force. Atomic nuclei are surrounded by negatively charged elementary particles called electrons, which have 1/1,836th the mass of the proton. Our galaxy is pervaded by high-speed protons, neutrons, electrons, and helium nuclei, as well as by larger atomic nuclei that have been stripped of their electrons

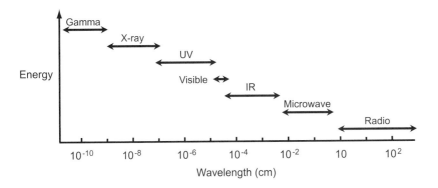

FIGURE 9.1. THE ELECTROMAGNETIC SPECTRUM.
The frequency and energy of electromagnetic radiation increase inversely with wavelength.

through collisions or by acceleration to velocities close to light speed. This ionizing radiation readily penetrates tissues and causes genetic mutations. It is therefore the main danger to life in space (Kiefer et al. 1994).

The types of ionizing radiation encountered just beyond our atmosphere have primarily one of three origins: solar cosmic radiation (SCR), galactic cosmic radiation (GCR), or geomagnetic radiation trapped in the Van Allen belts. The main features of these types of radiation are different.

GCR originates in the Milky Way outside the Solar System, where particles can be accelerated to relativistic velocities. These particles are roughly 98 percent baryons and 2 percent electrons. The baryons are primarily protons (87 percent), that is, hydrogen nuclei (atomic number or Z=1). Another 12 percent are alpha particles (or helium nuclei, Z=4), and the last 1 percent are heavy ions. The heavy ions include nuclei of all naturally occurring elements (up to and including uranium, Z=92). A tiny number of very high-energy particles are extragalactic and derive from ancient, distant supernovas and quasars.

SCR consists of low-energy particles and solar particle events (SPE) of the solar wind or heliosphere. SPE are sporadic bursts of highly energetic protons, helium ions, and heavy ions and electrons emitted by magnetic solar disturbances. SPE start rapidly and generally last a few hours, although proton events may last for a few days. Very rarely, the particle energy is high enough to sterilize Earth, were it not for our strong magnetic field (Baker et al. 2004).

GCRs enter the Solar System when the energy of the particles can overcome the outbound pressure of the solar wind, or heliosphere. The heliosphere is magnetic, and arriving GCRs are deflected and attenuated. The strength of the solar wind varies with the eleven-year solar cycle; hence GCR fluxes vary inversely with the solar cycle. This is called *solar modulation*. GCR flux peaks at the minimum solar activity and dips at high solar activity; the peak difference is about fivefold. A typical eleven-year solar cycle usually has four quiet years with an occasional SPE and seven active years with a large number of SPE near the solar maximum.

The Van Allen belts provide shielding against SPE and GCR by trapping incoming electrically charged particle radiation. These particles are deflected magnetically, and only at the poles is Earth's surface fully exposed. In contrast, at the magnetic equators only the most highly energetic particles ever hit the ground.

The atmosphere, because it is dense, provides Earth with another level of protection from cosmic radiation. The importance of the atmosphere was illustrated earlier by the increase in radiation with altitude. The radiation per unit of surface area doubles roughly every mile above sea level, up to about six miles. The amount of cosmic radiation received by the crew of a spacecraft also varies with the vehicle's trajectory and the length of the mission. As a result, mission planners carefully factor these exposures into the flight plan.

The long-term health effects of ionizing radiation are of greatest concern in staying in space, but nonionizing radiation is also harmful, as anyone knows who has had a sunburn, snow blindness, or skin cancer. With no atmosphere, solar ultraviolet (UV) radiation would produce blinding retinal burns in a matter of seconds. Your mother taught you never to look directly at the Sun for a good reason. Solar UV is also dangerous to the eye on the Moon and on Mars, but it is nonpenetrating, and shielding against it is simple.

THE ALARA PRINCIPLE

In deep space, we have three alternatives for dealing with cosmic radiation. The first is radiation shielding for spacecraft and astronauts, the second is biochemical countermeasures that improve tolerance to space radiation, and the third is the ALARA principle. ALARA allows a higher but "generally acceptable risk" of death from a cumulative lifetime dose of radiation. In other words, a few astronauts will die early of leukemia, lung cancer, and brain damage, but they chose to be astronauts, so it's on them.

These three alternatives are not mutually exclusive, but overtly life-shortening practices are objectionable. Waiting until the radiation exposures are acceptable is not the same as saying that they are not allowable. This also allows for sending older volunteers into space.

Ionizing radiation causes tissue damage in several ways and is best seen by shifting our attention from the wave to the photon, or particle, properties of light. As photons pass through tissues, they transfer energy in discrete packets. This energy transfer causes both dose-dependent and dose-independent types of damage. Dose effects relate to the apparent mass and velocity of the particles, which account for the health effects of heavy ions, baryons, and atomic nuclei. If a cell can be killed by a single fast, heavy ion passing through its nucleus, then ten heavy particles can kill ten cells, and so forth. This is dose dependence. There are also dose-independent effects, for which the probability of cell death or damage may be detected in a population but cannot be predicted precisely for an individual. These effects are therefore random, or *stochastic*, events.

Damage by ionizing radiation involves three processes: the *photoelectric effect*, the *Compton effect*, and *pair production*. Low-energy radiation produces a photoelectric effect when an arriving or *incident* photon interacts with an electron in the outer shell of a target atom. If the incident photon has more energy than the energy that binds the electron to the atom, the photon is absorbed, and the electron leaves with the energy of the photon, minus the binding energy.

At a higher energy, an incident photon interacting with an electron transfers part of its energy to the electron and continues on as a less energetic photon. This is the Compton effect. When a very-high-energy photon is absorbed by an atom, a positron and an electron are produced. This is pair production. A positron has the mass of an electron but a positive charge instead. Positrons and electrons rapidly annihilate each other, emitting two photons that fly away in opposite directions. Heavy particles can produce direct molecular damage as well as generate all three kinds of secondary photon events.

Our senses do not detect ionizing radiation or tell us that it is damaging our bodies, so we must rely on instruments to measure radioactivity. The standard international (SI) unit for an absorbed dose of ionizing radiation is the *gray* (Gy), which deposits 1 joule/kg of energy (also 100 rads). The Gy is multiplied by a biological quality factor for the type of radiation in order to compare radiation doses absorbed by the tissues, because different types of radiation produce different effects. This yields a unit called the *sievert* (Sv), and most exposures are recorded in thousandths of a sievert (millisieverts; mSv).

A little more on the biological quality factor: Heavy particles, like neutrons, produce more effects than X-rays at the same dose. This effect is standardized (to a 250 kV photon) to derive a relative biologic effectiveness (RBE); high RBE means a big effect. That is, ionizing particles have the highest RBE. For gamma rays and X-rays the RBE is one, while for neutrons the RBE can be as high as twenty. RBE also involves the ionization along the radiation track through the tissue. The energy deposited is called the linear energy transfer, or LET (in kilovolts per micron; $kV/10^{6}$ meter). High- and low-LET radiation have different biological effects, correlating roughly with the amount of O_2 present in the cell. High-LET radiation damages oxygenated and hypoxic cells in roughly the same way, but low-LET radiation causes less damage to hypoxic cells because it produces fewer reactive oxygen species (ROS).

Certain types of radiation interact with O_2 or with carbon, nitrogen, metals, or other atomic nuclei from which electrons can be stripped to produce free radicals. Free radicals contain unpaired electrons and are unstable, with lifespans of fractions of a second. In cells and tissues, radiation mainly generates ROS. For instance, X-rays generate the hydroxyl radical (·OH), the most damaging of all ROS. Wherever ·OH is generated, it immediately oxidizes the nearest molecule of protein, lipid, or nucleic acid.

A major site of radiation damage is the chromosome. Chromosomes contain chromatin, which encases in protective proteins the double-stranded molecules of DNA that carry our genetic information. Chromosomes self-replicate and provide progeny cells with the genetic information of the parents. DNA is an ideal propagator of life because it is stable, but it does have a capacity for tiny random variations (mutations) that promote evolutionary change.

DNA damage can take several chemical forms, as shown in figure 9.2. These include chemical additions (adducts), base deletions, and breaks in one or both strands. The protective proteins around DNA, such as histones, can also be damaged. Mammalian cells also have repair enzymes for many types of DNA damage, like single-strand breaks, and this is an important part of our natural defense against cancer and radiation. Radiation damage can lead to genomic instability through permanent genetic mutations or the modification of certain epigenetic marks on either DNA or proteins. These effects can interfere with the expression of tumor suppressor genes or with the DNA repair enzymes. If both DNA strands break, however, the molecule is difficult to repair, and most cells die at that point (Lobrich et al. 1995).

Less severe DNA damage may also be lethal but manifest only when the damaged cells divide. For instance, in mitosis, the cell precisely duplicates its chromosomes, and if any are broken, the broken ends can recombine or cross over with the ends of different chromosomes, often effectively. These events are sources of natural genetic diversity for the second type of cell division, meiosis, because they allow rearrangements or sorting of chromosomes in germ cells (ova and sperm). However, chromosomal crossovers in non-germ cells (somatic cells) are often lethal. If large numbers of cells in a tissue must divide rapidly, the tissue is especially vulnerable to radiation. This is also the basis for radiation's effect on cancer cells, but certain normal tissues,

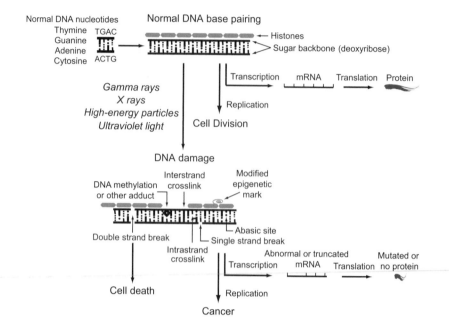

FIGURE 9.2. DIFFERENT TYPES OF MOLECULAR DNA DAMAGE.
Radiation produces many DNA chemical mutations and epigenetic changes in DNA bases and associated proteins. Some modifications can be repaired, some are passed through to daughter cells, and others are lethal. The top part of the diagram shows the normal processes required to synthesize new proteins; the bottom part shows the many effects of radiation.

such as the bone marrow, the lining of blood vessels (endothelial cells), and the outside (epithelial) cell lining of the intestine are highly sensitive to radiation damage.

The measurement of radiation doses in space is a tough problem. Biological dosimeters detect and record the amount and type of radiation received using different kinds of detectors; however, detectors near the body do not see the actual dose received by the organs. This shortcoming can be overcome by estimating the total radiation exposure from actual damage to chromosomes in the body, for instance, in white blood cells.

Damage to lymphocytes, for instance, can be compared to that of control lymphocytes exposed to known doses of radiation. This type of analysis can detect whole-body radiation doses that are very low (about 0.1 Sv). It captures a functional effect of radiation on the white blood cell that can be used to help us understand both the radiation exposure history of the person and the combined effects of radiation and other biological effects of space on groups of people.

The radiation sensitivity of lymphocytes has even been considered as a test to screen astronauts for natural radiation resistance. Screening strategies based on physical attributes such as height or vision always trouble the populist, but selecting space explorers on the basis of whether their lymphocytes survive a death ray in a dish would exasperate otherwise highly qualified candidates whose lymphocytes were not up to snuff. This simply puts too much weight on the behavior of a single type of cell in a simulation.

Here is an analogous case that illustrates a frequent flaw in the logic of such methods of screening. A small heart anomaly is present in about 25 percent of healthy people, an anomaly called a *patent foramen ovale*, or PFO. A PFO is a small passage between the two upper chambers of heart, the atria. On rare occasions, it can cause a problem by allowing a blood clot to pass from the right side, or venous circulation, to the left side, or arterial circulation, without being filtered by the lungs. This clot can then migrate to an artery to the brain, causing a stroke. This is called paradoxical embolization.

Not infrequently, scuba divers during ascent or astronauts decompressing in spacesuits before EVA develop bubbles in their venous blood. Such bubbles are easily filtered by the lungs and normally leave the body harmlessly. Occasionally, the load of gas leaving solution is too large and causes decompression sickness, or the bends. If this happens with a PFO present, venous bubbles can cross the heart into the arterial circulation, causing paradoxical gas embolization.

A PFO is easily detected using noninvasive medical ultrasound of the heart or echocardiography. Some years ago, someone had the brilliant idea of screening astronauts with ultrasound and excluding those with PFOs from EVA in order to avoid this problem. Predictably, astronauts protested because people with PFOs dive routinely with no increased risk of DCS. DCS is caused by events related to the supersaturated gas burden in the body (which causes the bubbles during ascent), not by the PFO itself. If someone with a PFO does not properly decompress, the ensuing DCS may be complicated by paradoxical embolism, but this is altogether a consequence of improper decompression. The furor finally died down when NASA decided to institute safer decompression procedures rather than screen astronauts for PFOs.

The argument against screening healthy candidates for sensitivity to radiation with a blood test is different than using the blood test to detect existing radiation damage, for instance, from persistent chromosomal abnormalities. This area of research is in its infancy, but even if the prior and acute effects of radiation are known, the late effects are hard to predict. The worst effect, the induction of cancer, may not occur for twenty or more years after the exposure. In an exposed population, the number of cancer cases increases over time, but the events are stochastic, and ultimately, a small percentage of the group is affected. This means that not everyone with the same radiation dose gets cancer, even people of the same age, gender, and overall health.

This difference derives mainly from three factors. The first is individual differences in radiation tolerance, including the ability to repair damage. The second is that mutations that transform cells into cancer are random, and not everyone gets the same ones. Finally, as people age, they die of other causes. Older people, primarily those over the age of fifty-five, whose life expectancy is twenty or so years, thus have a lower risk of radiation-induced cancer than younger people. In other words, as life expectancy becomes shorter, the latent period for cancer development is longer than the host's expected lifespan.

These aspects of radiation are a concern for spaceflight, but radiation tolerance is also a factor. Tolerance involves the cell's recovery and repair mechanisms and is affected by how the radiation dose is received. A single large dose or fraction of radiation given over hours has a far greater effect than the same dose given over days, weeks, or longer. And despite recovery from acute radiation injury, late effects may still develop.

The extent of recovery after radiation damage is related to the number of progenitor or *stem cells* that remain alive in the bone marrow and other organs. Stem cells are essentially blank (undifferentiated) cells produced in both embryonic and adult varieties that can give rise to essentially all types of specialized cells. If too many stem cells are destroyed and cannot be replaced from healthy ones, radiation injury will be more severe and persistent.

Radiation impairs survival in all types of mammalian cells, and radiation dose is related in a predictable way to the fraction of surviving cells. For low doses of X-rays or gamma rays, a survival curve may indicate that the cells are repairing a certain amount of radiation damage. For strongly ionizing radiation, however, the repair effect disappears, and death depends strictly on dose.

Radiation therapy in cancer works because radiation is given in timed increments and in doses, called fractions, that preferentially kill rapidly dividing tumor cells over normal cells (Fornace et al. 2000). Some tumor cells are naturally resistant to radiation because of innate tolerance factors. In some cases, these factors promote hypoxia and increase the expression of tumor protection genes known as *oncogenes*.

Cells also produce a range of natural molecules in response to radiation, including hormones, growth factors, antioxidant enzymes, and small signal molecules called cytokines. Cytokines are involved mainly in the cell-to-cell regulation of the body's inflammatory response. Radiation induces some cytokines that promote inflammation, such as tumor necrosis factor alpha (TNF) and interleukin-1 (IL-1). These two cytokines were originally identified because they cause wasting in cancer patients and fever during infections, respectively. They help fight infection and protect blood cells from radiation, but in some cases, TNF enhances the ability of radiation to kill cancer cells. This contrast illustrates the complexity of radiation effects but does suggest that better strategies can be worked out for radiation protection.

The production of growth factor molecules is an important radiation repair mechanism. Different growth factors, including fibroblast growth factor (FGF), platelet-derived growth factor (PDGF), vascular endothelial growth factor (VEGF), and transforming growth factor (TGF)-beta, play roles in the repair of radiation injury, but most do not improve radiation tolerance. Some actually perpetuate the influx of inflammatory cells into blood vessels, causing inflammation and worsening cellular damage via "bystander" effects. TGF-beta promotes repair but also causes inflammation and extensive scarring or fibrosis after radiation, especially in the lungs and the skin.

RADIATION'S EFFECTS

Much of our knowledge of the mass effects of radiation comes from the survivors of Hiroshima and Nagasaki, Pacific islanders exposed to fallout from U.S. thermonuclear bomb tests, and rare reactor accidents like Chernobyl. These calamitous events confirm both immediate and late deaths from radiation (Schull 1998). The dose and type of radiation and radiation tolerance are all important. In survivors of the atomic bomb blasts, late carcinogenic and genetic effects correlated with the distance from the explosion's epicenter. The development of chromosome breakage and cancer was greatest in survivors within about a mile (2 km) of the epicenter (Schull 1998, Preston et al. 2003). The young and the elderly are most sensitive, and women tolerate radiation better than men. Exposure to less than two Gy generally requires no treatment, but exposure to 3.25 Gy without treatment kills about half the people.

There are three well-recognized radiation sickness syndromes. The cerebrovascular syndrome (twenty to one hundred Gy) damages the brain and blood vessels and causes death within forty-eight hours. It kills so quickly that there is no time to see the damage to other body systems. At ten to twenty Gy, the sloughing of intestinal cells leads to gastrointestinal syndrome, causing diarrhea and dehydration, followed in about ten days by infection and death. Patients with the gastrointestinal syndrome may live for weeks or months and may survive if they receive fluids, electrolytes, blood products, and antibiotics. The hematopoietic syndrome is caused by bone marrow failure after exposure to two to eight Gy of radiation. It takes a few weeks to develop because mature cells in the bone marrow must become depleted while new ones are no longer being produced. Lymphocytes die first, followed by other white cells and megakaryocytes, the blood platelet precursor. Death usually occurs from bleeding and infection before anemia appears.

The effects of radioactive fallout are not fully understood. The analysis of radiation exposures from the Marshall Islands and the 1986 Chernobyl accident has been invaluable, but the numbers are small. In time, Chernobyl will cause 6,000 to 10,000 excess deaths, and at least eight types of cancer are seen among survivors, but especially important are leukemia and cancers of the thyroid, breast, colon, and lung.

In 2002, the first true estimates of the number of cancer deaths from the atmospheric nuclear test programs of 1951 to 1962 appeared in a report prepared for Congress by the Department of Health and Human Services. Over this period, 390 nuclear bombs were exploded above ground—mostly by the United States and the Soviet Union (Simon et al. 2006). Radioactive fallout encircled the Earth.

It was assumed that these exposures would not cause many cancers, but the excess thyroid cancer and leukemia cases turned out to be far greater than most thought possible. Between 1951 and 2000, some 11,000 more people died from these two cancers in the United States alone, particularly in those exposed at a very young age. Thousands have died in other countries too, especially around the infamous Soviet Semipalantinsk test site.

Astronauts usually do not face these types of radiation, and the quality and consequences of cosmic radiation are different (Durante and Cucinotta 2008). However, these experiences warn us not to underestimate the risks.

We lack reliable risk estimates of the radiation doses that astronauts will receive after they leave the Earth's atmosphere and Van Allen belts, but radiation will be a serious health risk on the Moon and Mars, especially at places without atmospheres or near Jupiter (Cucinotta et al. 2005).

There are two problems: the constant exposure to solar and galactic radiation and the sporadic exposure to high-level SPE and galactic bursts. Since most of the radiation is solar, you would think that moving away from the Sun would decrease the risk, but the solar wind protects the Solar System from GCR. It repels GCR, and thus the amount of galactic radiation reaching us varies inversely with our distance from the Sun. Thus Mars, even though it is farther from the Sun, is not safe. Mars orbiter data indicate that radiation levels there are 2.5 times those on the ISS because of Mars's thin atmosphere and lack of a magnetic field. The surface of Mars is safer than deep space (NASA 1999), but astronauts will still need better radiation shielding than is currently available.

Radiation exposure limits are especially hard to set when the level of risk is poorly established (Cucinotta et al. 2004). This uncertainty has to do mainly with unknowns in biological dose, individual susceptibility, and the prediction of late effects such as cancer. For instance, astronauts are exposed to ionizing radiation from GCR and SPE at different energies and on different profiles. Even if the dose estimate is precise, it is difficult to predict late effects that could be attributable to natural disease and other causes of mortality instead.

Scientists have known about this uncertainty for years, but only recently has it been evaluated carefully. Cancer risk assessments are being done to determine what kind of data would increase the level of confidence in risk predictions. At this point, the expected lifetime cancer risk for astronauts after a three-year Mars mission is not predictable to within a factor of five. This means that risk calculations are less useful in planning protection for the crew.

In deep space, the late effects of radiation, including cancer and brain effects, will derive mainly from high-energy (HZE) ions in SPE and GCR. Because HZE effects will dominate, risk estimates from data from missions in LEO is unreliable. LEO is comparatively safe because the HZE is deflected by the Van Allen belts; hence, little HZE exposure data exist. Only during the Apollo era, with its missions beyond the Van Allen belts, were radiation doses dominated by GCR. This is important because HZE is not well attenuated by conventional shielding, like aluminum, or by current spacesuits. Stopping these particles produces secondary radiation from atomic interactions in solid materials. These phenomena are called *Bremsstrahlung* and *spallation* (Cucinotta 2002, 2005).

Bremsstrahlung occurs when high-speed electrons slow down inside certain materials, emitting X-rays as a result. Spallation is the generation of particles from high-energy interactions with atomic nuclei. GCR and SPE contain high-velocity protons that penetrate most shielding and cause tissue damage from both primary and secondary effects. Without proper shelter, high-energy SPE can cause fatal radiation sickness.

The Van Allen belts trap electrons with energies of a few MeV, electrons that could otherwise penetrate two centimeters of H_2O or give off Bremsstrahlung that could penetrate even more deeply. Trapped protons have energies up to several hundred MeV, but they usually penetrate less than one centimeter of H_2O. Standard shielding thus effectively attenuates trapped radiation at low energy. The high-energy component is small but easily penetrates spacecraft and tissues and produces secondary effects (Cucinotta 2002).

In contrast, GCR exposure is dominated in LEO by high-energy particles because the Van Allen belts block the low energies. The remaining particles travel at relativistic velocities and undergo multiple atomic interactions within shielding and other materials, including tissues. This produces a surge of secondary particles, including neutrons, protons, helium, and heavy ions, especially in materials of high atomic number. In fact, astronauts with their eyes closed report bright scintillations, called *phosphenes*, which are thought to be related to high-speed heavy ions.

The origin of phosphenes was first attributed to charged particles passing though the eye and either hitting retinal cells or causing flashes of blue Cerenkov radiation by disrupting the electromagnetic properties of the vitreous humor of the eye. They were thought to be insignificant, but later work raised the possibility that fast particles are hitting the brain in the optical corona or perhaps in the visual cortex (Fuglesang et al. 2006). If true, this could have major implications for vision and other brain functions on long missions in deep space.

The best shielding against secondary radiation from GCR would be hydrogen-rich materials of low atomic mass, like water. These materials limit secondary particle production and are quite effective per unit of mass in stopping heavy ions after atomic collisions. This was shown to work by shuttle experiments comparing polyethylene $(HCH)_n$ and aluminum spheres of an equivalent size and mass.

In the past, NASA had higher radiation exposure limits for astronauts than are allowed terrestrial radiation workers because they flew a limited number of missions. The nuclear industry has age-specific, career dose limits computed as age x 0.01 Sv. To avoid large lifetime doses, U.S. radiation workers have yearly exposure limits of fifty mSv. Terrestrial radiation workers rarely approach their lifetime limits. NASA's system for astronauts is similar, and they are considered radiation workers (Cucinotta 2002, 2005).

NASA accepts space radiation exposures that are associated with a 3 percent increase in the lifetime risk of developing cancer. The cancer risk is superimposed on the expected "normal" cancer rates in a population of the astronaut's age and gender. The incidence of nearly all cancers increases with age, reflecting natural genomic instability and the accumulation of environmental chemical effects and spontaneous mutations. Thus, the ratio of radiation-induced to normal cancer risk is less for older than for younger astronauts.

The level of uncertainty in these effects is high, making it hard to estimate an actual statistical confidence interval. It is easy to see the importance of this uncertainty by plotting the risk of cancer as a function of the linear energy transfer (LET) across a range of energies. This has been done in figure 9.3. The prediction varies with LET, and the statistical confidence interval, as a measure of variability, is highest for radiation doses associated with the highest probability of developing cancer.

When statistical confidence is low, it is difficult to evaluate risk-mitigating strategies for spaceflight radiation apart from improving the radiation shielding. But the cost and stability of every new shielding strategy must also be compared against proven shielding strategies. For instance, water is excellent shielding for neutrons, but it is heavy and expensive to launch in the swimming-pool quantities necessary to block cosmic neutrons. Another strategy is to limit the time in deep space by investing in faster propulsion. And biological countermeasures are also ripe for improvement.

Some NASA scientists think that better radiation risk predictions would increase the number of permissible days in space simply by improving the confidence of the estimate. In other words, the uncertainty is so high that better predictions would probably increase the astronaut's allowable time in space. But new information might just as well increase the predicted risk, and odd effects may be encountered too, such as a lower cancer risk for certain radiation exposures. Indeed, some populations that receive slightly more than average natural background radiation actually show a small reduction in cancer risk—an effect called *hormesis*.

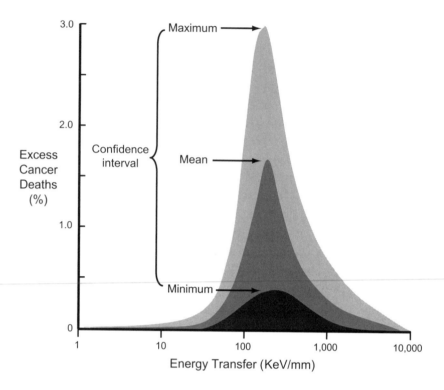

FIGURE 9.3. THE HIGH UNCERTAINTY IN THE BIOLOGICAL EFFECTS OF COSMIC RADIATION.
The graph illustrates the high variability (wide 95 percent confidence intervals) in the predicted probability of excess cancer deaths at linear energy transfer rates relevant to exposure to cosmic radiation.
Source: Adapted from data in Cucinotta et al. (2004).

COUNTERMEASURES

The science of radiation-induced damage should eventually lead to better biological protection for space radiation. Highly effective countermeasures would be groundbreaking, but they must work for high-LET radiation for long periods with few side effects. Unique protection strategies for relativistic-speed HZE ions will require independent validation because agents that protect against low-LET radiation may not protect against high-speed heavy particles.

The United States has developed drugs for protection against atomic blasts, to reduce radiation therapy side effects, and more recently for counterterrorism. Many of the biodefense agents are free radical scavengers or antioxidants (Wan 2006). Other drugs stimulate the recovery of stem cell populations in sensitive tissues, which might block or limit acute damage.

Most protective drugs for radiation sickness have serious side effects. In some cases, combining lower doses of protective agents with natural antioxidants such as vitamins A and E helps alleviate the side effects, but cellular antioxidant levels are unlikely to become high enough to prevent DNA damage from acute high-LET radiation. On the other hand, antioxidants may offer some benefit against the late effects (Fornace et al. 2000).

Drugs that prevent the growth of premalignant cells in radiation-exposed tissues are also being explored. The estrogen antagonist tamoxifen, used against breast cancer, for instance, may block tumor induction caused by several types of radiation. Finally, the age and gender dependence of cancer suggests avenues for crew selection to reduce the cancer risk. Of course, the notion of sending only postmenopausal women to Mars would be rather contentious!

There are breakthroughs in radioprotection on the horizon that are in need of research. Aspects of genetic engineering and stem cell biology are especially promising if cells can be endowed permanently with radiation resistance through the expression of biological factors that confer resistance before the exposure. This would reduce the amount of damage that must be repaired during and after the exposure. The ability to replace radiation-damaged cells with fresh stem cells has also attracted attention. Radiation resistance is feasible: experiments of nature indicate that certain organisms, especially bacteria, survive doses of ionizing radiation thousands of times greater than those that would kill people.

One such bacterium is the extremophile *Deinococcus radiodurans*, named for its radiation resistance. This resistance is not related to fewer DNA double-strand breaks from radiation exposures, which develop at a normal rate, but rather to its active DNA repair enzymes, which can repair double-strand breaks and are resistant to radiation damage. This resistance has been traced to an unusually high intracellular ratio of manganese to iron (Mn/Fe), which somehow protects the proteins from free radical damage. Thus this microbe has developed tolerance by evolving proteins for a DNA repair system that are resistant to radiation damage (Daly et al. 2007). This suggests that natural ways around DNA radiation damage may be inducible in other living organisms or perhaps given to people, so that more space radiation can be tolerated with fewer adverse health effects.

10. TINY BUBBLES

You might think that by now we could have developed a coherent plan for human space travel, but the lack of information on the hazards of life outside the Van Allen belts, where only a handful of Apollo astronauts have ever been, means that all bets are off. The longer-term effect of radiation on the human body is still a big unknown, and sound statistical thinking points out weaknesses in our understanding rather than a solid estimate of risk. Other unknowns must also be taken into account in putting together a sequence of steps for space exploration. The rationale for a lunar base was set out earlier, but in order to get from the Moon to Mars, both transportation and surface technologies will require new advances.

DESTINATION MARS

To illustrate the challenges of interplanetary exploration, let's examine solutions to a couple of key problems raised by an attempt to travel to Mars. The leaders in the field are well aware of these problems, but they have not agreed upon the solutions, mainly because of a lack of information. Assuming a moonbase proves practicable, the surface technologies can be adapted for Mars while the transportation issues are being worked out, issues such as getting off the planet safely after the mission.

Interplanetary missions are risk multipliers; they superimpose *getting there* and *staying there* in a unique way. Time and distance increase the risk of a lethal failure occurring, one for which corrective action is not possible in deep space. The longer you spend on the planet, the more time there is for the surface technology to fail and prevent you from getting home. Let's examine the risks and see if we can minimize them. Doing so might be costly, but we'll let someone else worry about that.

NASA's Space and Life Sciences Directorate has gone through the trouble of tallying and stratifying the risks of deep space exploration, including a Mars mission. More than one hundred risk factors have been identified, and most are attributed to a relatively small group of factors. The Pareto principle is at work: 20 percent of the causes account for 80 percent of the risk. Accordingly, I have focused on the top 20 percent of the risk factors.

These factors have less to do with exploding boosters, disintegrating heat shields, or wayward egress than with breaches in life support, radiation shielding, and illness. Bone and muscle loss and the psychosocial aspects of living in a tightly confined space for a long time also come into play. Although group conflicts are highly disruptive, these are best avoided by meticulous crew selection and training and are therefore set aside here. You may have heard of the Mars500 experiment, which ended in 2011: to study the psychological effects of living in the sort of confined environment that would have to be endured during the long journey to Mars, six men volunteered to live in a seventy-two-square-meter cabin in Moscow for 520 days, and they did so peaceably.

The hope of sending astronauts to Mars using an upgraded version of our new space transportation system ultimately depends both on its cost and on how much experience we have with the system. A round trip to Mars is roughly the same distance a thousand Moon trips—it is too far away and too much of a one-shot deal to squander big resources without expectation of a big return. But what is a reasonable return on a Mars investment? I cannot answer that question directly, but perhaps there is another way to look at it.

After the Moon, and apart from an asteroid, Mars or its moons are the only options. Asteroid missions, which are not on par with planetary missions, are left for later. Earlier, we also considered Venus, which is closer than Mars—but also closer to the Sun. We decided against it because it is hot enough to melt lead. It is bad luck that Venus turned out to be so inhospitable. Thus,

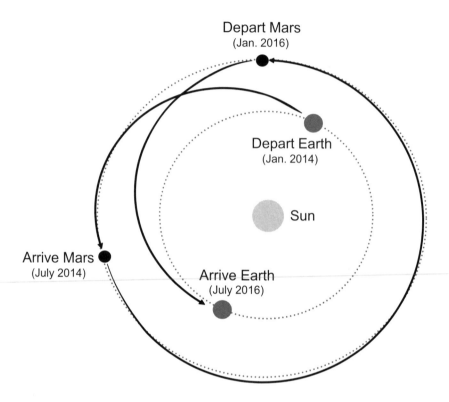

FIGURE 10.1. THE 1998 NASA MANNED MARS REFERENCE MISSION ORIGINALLY PROPOSED FOR 2014 THROUGH 2016.

our first interplanetary foray will be to Mars or one of its moons. The question is: "when?" Planning for a Mars mission is already quite advanced; in fact, the text of the 1998 NASA Mars Reference mission (Drake 1998, NASA 1999, Charles 1999), devised originally to evaluate the feasibility of the mission in 2014, is rather dog-eared (figure 10.1) at this point.

Although it is now over ten years old, the plan is still informative, for two reasons. The first is that the fifteen-year window implied by the reference mission has gotten longer, not shorter, as we have learned more about the issues. The second is its advocacy of the so-called fast or conjunction mission, consisting of six months in deep space each way, returning 2.5 years later, after the astronauts spend eighteen months on Mars. This is still the preferred format because it minimizes the time in transit and maximizes the time on the ground. The alternative, an opposition mission, requires more time in space and less time on Mars.

Because Mars has a thin atmosphere, astronauts would receive less cosmic radiation by spending more time on the planet. After six months in transit, they should arrive in shape to operate in a spacesuit at 38 percent Earth gravity and healthy enough that when they leave for home eighteen months later, they could survive six more months in space. At least that's the plan.

Because sending resupply ships to the surface of Mars is not going to be feasible for a long time, the systems and cargo on the original ship or ships must support the entire mission until Mars surface technologies can provide additional resources. For logistical and safety reasons, a working habitat as proposed by Robert Zubrin and others would be set up on the Red Planet some time before a crew ship actually left Earth.

An established base ensures that power, atmosphere, water, and food for the crew is waiting when they arrive. This habitat must be maintained from Earth while the crew is en route—a good reason not to leave it unattended for too long. And the ship must have contingencies to deal with failures. This is why proponents of advanced propulsion systems that could carry astronauts to Mars in under sixty days think it is the key to Mars.

If a Mars habitat is set up remotely, a dual approach to the renewable atmosphere makes sense: a "disposable" temporary atmosphere and a regenerative cycle that replenishes the atmosphere with the help of green plants. Although plants may also be grown en route, most experts favor gardening on Mars, where there is decent gravity (NRC 1998). Moreover, gardening on Mars is feasible with plants that don't require genetic alterations. They must simply be selected and bred for hardiness, productivity, and fecundity in an appropriate artificial soil.

The reason plants still cannot be trusted to provide the bulk of food and O_2 on a Mars spacecraft has to do with the carbon cycle. Recall that on Earth, sunlight and H_2O plus CO_2 from animal respiration (or factories) is coupled to the release of O_2 by plants performing oxygenic photosynthesis. Plants use CO_2 (and get nitrogen from waste products or fertilizers), and animals eat plants. Dead material is recycled primarily by saprophytic bacteria in the soil. To complete the cycle, terrestrial animals, including human beings, breathe in O_2 and exhale CO_2.

This cycle can be moved into space, by using artificial lighting and plants to recycle the atmosphere on a spacecraft, but compared with an upgraded SLS such a craft would be very large and very expensive. The heavier it is, the more it will cost to operate at high velocity, and the slower it is, the greater the microgravity and cosmic ray problems.

Once bioregenerative surface technology is placed on Mars, adding a crew becomes easier. Such a system would be neither a simple greenhouse nor an Earthlike ecosystem (Sridhar et al. 2000) because it would not have huge natural holding areas like those that store and recycle waste here on Earth. Our holding areas include the atmosphere, oceans, and continental land masses. No one really knows how small those areas can be and still provide adequate buffering for stable and efficient bioregenerative systems. Bioregenerative experiments have failed dismally on Earth, and horticulture has another downside: it increases the reservoir of niches for microorganisms, including dangerous bacteria and fungi.

MICROBES IN SPACE

On November 8, 2011, the Russian Space Agency launched a 170-million-dollar, fourteen-ton spacecraft on a three-year mission to land on Phobos, collect some soil, and then return a small capsule of samples to Earth. The *Phobos-Grunt* spacecraft was also carrying a Chinese orbiter to monitor Mars dust storms and a study funded by the Planetary Society called LIFE (Living Interplanetary Life Experiment) intended to test the survival of Archaea, bacteria, and tiny eukaryotes in deep space. Among the ten little organisms, the passenger considered most likely to survive the sea of radiation was our old friend *Deinococcus radiodurans*.

The spacecraft lugged eight tons of fuel into LEO, but the rocket to send it to Mars failed to fire. The ship stopped communicating with its managers and missed its last call for the Red Planet. The orbit gradually decayed, the chief engineer apologized profusely, and the craft unceremoniously burned up over the Pacific Ocean, off the coast of Chile, two months later. Although for Russia the mission was yet another great indignity for Mars spacecraft, the furor it ignited about the wisdom of shooting unattended microorganisms at Mars momentarily died down when the thing crashed.

If you think carefully about the LIFE experiment, you will see only one question it could have answered: which (if any) of these ten organisms can survive a nearly naked trip into deep space? The answer sheds not one scintilla of light on whether life actually originated on Earth—the principal reason given for the experiment. The architects of LIFE were wondering whether life came here from elsewhere, a theory about a "cosmic teapot" called *panspermia*.

Many microbes are quite resistant to harsh environments, and some thrive under remarkably extreme conditions. These extremophiles include *D. radiodurans*; there are also thermophiles (heat), halophiles (high salt), barophiles (high pressure), xerophiles (dry), cryophiles (cold), acidophiles (acid), and others. These little darlings of exobiology prove that life can handle a wide array of physical environments, and most extremophiles become dormant in hard times, forming spores until conditions for growth improve.

Fastidious microbes are often associated with terrestrial plants or animals, and some cause diseases (these are pathogens). The ancient bacteria of genus *Clostridium*, for instance, grow only by fermenting carbon compounds in the absence of oxygen. These strict anaerobes are natural saprophytes and play important roles in the carbon cycle, but they also cause tetanus, botulism, gas gangrene, antibiotic-associated diarrhea, and other diseases. When exposed to air, they form endospores that remain dormant for months or years.

The possibility that such organisms might stow away on interplanetary spacecraft is well appreciated. Some years ago, an editor of *Discover* magazine raised the possibility that the Mars rovers may have accidentally carried a common skin bacterium, *Bacillus safensis*, to Mars from the Jet Propulsion Laboratory in Pasadena. The Mars Science Laboratory and *Curiosity* rover are taking another look for signs of past life on the Red Planet. Hopefully, *Curiosity* did not harbor some curious Earth microbe, which would confuse the issue about whether life once arose indigenously on our neighboring planet. NASA goes to great lengths to avoid such confounds, and it is not clear that even spore-forming bacteria could survive in interplanetary space or with the level of UV on Mars. However, one mistake can create a permanent mess.

Once people get involved, microbes will go to Mars no matter what we do. Billions of bacteria live in our pores and hair follicles; others peacefully luxuriate in our intestines. There is no practical way, antiseptics and antibiotics included, to eliminate them completely. Moreover, a Mars ship carrying plants gives microbes innumerable places to hide.

In the early days of spaceflight, Russian scientists detected microorganisms growing in strange places on their spaceships, such as in crevices in the structural materials of cabin interiors and equipment. These beasties were busily degrading the structure of spacecraft components. Russian scientists identified nearly one hundred different species of microbes in *Mir*'s cabin environment, including bacteria, molds, and fungi. A few species were potential pathogens, but most simply ate synthetic materials, including plastics.

Although microbial bioreactors, as noted earlier, do have certain attractive features, the other side of the coin is corrosion. Some microbes cause *biofouling*, especially those that make protective *biofilms* under stagnant or O_2-poor conditions. Biocorrosion takes different forms on structures exposed to the atmosphere and may involve filters, tubing, pumps, membranes, and water lines. This happens on all manned spacecraft, including the ISS. There are many types of biofouling, but the process often leaves mineral deposits and pits beneath living colonies. The pitting can be controlled by biocides and some conventional corrosion-control methods, for instance, by adding silver ions to cooling and water-circulating systems; however, over long periods, biocides are degraded or deposited on internal surfaces, releasing the microorganisms to grow (Roman and Wieland 2005).

Unchecked microbial growth in life-support systems hastens the deterioration of critical metal or polymer components but may also affect human health if pathogens escape into the atmosphere (LaDuc et al. 2004). This is a classic problem of source control in infectious diseases. Outbreaks of pneumonia by bacteria that thrive in wet environments, such as *Legionella pneumophilia* (Legionnaire's disease) and *pseudomonas aerugenosa*, are well known. On a Mars mission, someone would be responsible for monitoring and controlling the growth of microorganisms, especially dangerous ones, but it is hard to imagine how simple quarantine measures and biocides could prevent the contamination of Mars. This objection has been presented by some as a reason for not sending people to Mars at all.

The effect of microgravity on microbial virulence was studied in the pathogenic bacterium *Salmonella typhimurium*, grown aboard the shuttle STS-115 mission in 2006 (Wilson et al. 2007). *Salmonella typhimurium* is an enteric (intestinal) pathogen that causes gastroenteritis, and it is related to the strain of *Salmonella* that causes typhoid fever. *Salmonellosis* also causes fever and abdominal cramps, diarrhea, and dehydration for up to a week, and it may require hospitalization. The organism's natural reservoirs include livestock, poultry, and pet turtles, and it often causes outbreaks of diarrhea associated with eating lettuce, tomatoes, cantaloupe, other fresh produce, raw eggs, and chicken.

Salmonella gene expression in spaceflight was compared with ground controls, and virulence was checked by inoculating flown and unflown bacteria into mice. Over nine days in space, the microbe's gene expression changed, and it developed unique attributes, including the accumulation of an unusual external coat or matrix connected with the formation of biofilms. The space-flown microbes were significantly more virulent in mice, and after a standard inoculation, the lethality increased by about 50 percent.

An analysis indicated that 167 genes and seventy-three proteins had changed—sixty-nine genes were upregulated, and ninety-six were downregulated. These responses appear to be regulated by an RNA-binding protein called Hfq, which is activated when the bacterium senses shear stress in its environment. Although this microbe uses biofilms in its own defense, it is unclear exactly how this response increases its virulence. The idea that microgravity provides an environmental cue to induce a bacterial pathogen to increase its virulence has troublesome implications for long space missions, where a weakened immune response in people in a closed environment may increase the risk of life-threatening infections. This sounds like another good grant proposal for NASA.

GAS LEAKS

Since the verdict on microbes is still out, let's return to some unfinished business on the conservation of the respiratory gases, O_2 and CO_2. Gases escape all spacecraft at a slow rate. On the ISS, O_2 is generated onboard, CO_2 is dumped overboard, and nitrogen is stored as a "makeup" gas. The ISS is not perfectly sealed, and there are leaks to the outside, where the pressure is essentially zero. The leak rate is 1 percent per month, so if this gas was not made up, the atmosphere of the ISS would climb to the altitude of Mt. Everest in nine years and to the edge of space, the Karman line, in about forty years.

These leaks are no big deal because the gas escapes at a snail's pace, but they do bring to mind the Aristotelian dictum of *horror vacuii*—Nature abhors a vacuum. *Horror vacuii* is incorrect, of course; in nonquantum physics, space is a perfect vacuum. Earth's atmosphere is not sucked into space because it has mass and because our planet's gravity is enough to hold it. In the seventeenth century, the mass of the atmosphere led Torricelli to discover barometric pressure.

Let's think about O_2 for the crew of a Mars spacecraft. Recall that NASA allows for an astronaut to use about 0.9 kg of O_2 and generate about 1.05 kg of CO_2 per day. It might seem that the astronaut is producing more CO_2 than he or she is consuming O_2, but since CO_2 is heavier than O_2, this is not the case. At standard conditions, one mole of CO_2 weighs forty-four grams, while one mole of O_2 weighs thirty-two grams. Thus, an astronaut produces about twenty-four moles of CO_2 and consumes twenty-eight moles of O_2 each day. The ratio is 0.86, or about midrange for the respiratory quotient (RQ).

For a crew of six and a mission duration of three years, the mass of O_2 that must be carried is nearly six metric tons (5,913 kg), excluding a safety factor and the weight of the storage canisters. During the mission, nearly seven metric tons of CO_2 must be recycled or dumped overboard. If we take twice as much O_2 as we need and allow two tons for the storage weight, the ship would leave Earth with fourteen tons of O_2 in storage (30,800 lbs). To keep this in perspective, the payload capacity of the Space Shuttle was twenty-five metric tons (55,250 lb).

If all goes well, the ship would return from Mars with six tons of O_2. If some of the O_2 in metabolic CO_2 can be recycled, the ship could carry a smaller O_2 supply. If a bit over two-thirds of this CO_2 is recovered as O_2, about one-third of the original six tons of O_2 would be left after three years. This means the safety factor could be reduced to four tons and the ship would still return from Mars with six tons of O_2 in reserve.

CO_2 on the ISS is removed by the Sabatier reaction: $CO_2 + H_2 \rightarrow H_2O + CH_4$. Hydrogen ($H_2$) comes from the electrolysis of H_2O used to generate O_2, and the methane (CH_4) is dumped overboard. The water can be reused to regenerate O_2, but the conversion efficiency of CO_2 to O_2 on the ISS is only about 3 percent. In other words, the useful recycling of CO_2 to O_2 for a Mars mission is going to take much more efficient technology.

Apart from emphasizing the mass of cargo needed for Mars, this thinking introduces the idea of a safety factor. Using ISS data and a NASA nine-hundred-day mission protocol, we can also estimate the amounts of N_2 and O_2 gas lost from a Mars ship. Let's assume we breathe sea-level air, and we split the mission into three hundred days in space and six hundred days on Mars. This is less time on the planet than usual, but it makes the arithmetic easier. If the constant air leak rate is 0.111 kg per day for 900 days, the ship loses 100 kg of gas: 21 kg of O_2 and 79 kg of N_2. On Mars, the airlock loses 2 kg per day for 600 days, or 1,200 kg: 252 kg of O_2 and 948 kg of N_2. The net O_2 loss is 273 kg. Also notice that nearly five times as much N_2 make-up gas is needed as O_2.

If we start with twelve tons of O_2, this loss is 2.275 percent of the total. If we use high recycling to save on mass and start with only six tons of O_2, then the 273 kg of O_2 lost is 4.55 percent of the total. If recycling is very high and we start with two tons of O_2, then the leak takes 13.7 percent of the total.

A constant leak thus reduces the safety factor most when the recycling efficiency is highest.

THE AGE OF THE ASTRONAUT

You have become well acquainted with the medical problems of space, but ultimately none of them should prevent people from going to Mars if new space transportation systems reduce the radiation exposure and shorten the time in space. The big challenges will be landing, living there for eighteen months, and getting off the planet again.

Barring a catastrophe, radiation is the greatest health risk on a Mars mission, and without better shielding, a Mars crew would be exposed to cancer-causing levels of radiation. For NASA's 3 percent lifetime excess risk of fatal cancer, the decisive statistics are set not only by the cumulative dose but by the astronaut's age and gender at the start of the mission.

The relationship of cancer risk to age and gender means that older male and female astronauts can receive more radiation than younger ones. The career radiation exposure limits for fifty-five-year-old astronauts is two to three times greater than for twenty-five to forty-five-year-old astronauts. This implies older astronauts are better candidates for Mars travel than younger ones, at least insofar as radiation-induced cancer is concerned. By having lived longer, older astronauts also have a higher risk of age-related diseases, including cardiovascular and neurological disease and osteoporosis. However, radiation may also interact with these comorbid conditions as they progress. Thus, it is not clear whether it is better to send thirty-five-year-olds or fifty-five-year-olds to Mars.

By NASA's 3 percent rule, astronauts on a nine-hundred-day conjunction mission would barely get to Mars even if four inches (ten cm) of water is used to shield the spacecraft. A fifty-five-year-old man on a ship with standard shielding would reach the 3 percent limit during transit, while an extra ten cm of H_2O shielding would buy him some time on the planet—about half the mission. After returning, his *excess* cancer risk would be approximately 6 percent above the normal roughly 25 percent in his remaining life expectancy. Most people would brush off such a risk to be among the first to go Mars, but there are also unknowns about how the radiation interacts with other medical factors.

The muscle and bone loss on long spaceflights are not fatal, but the post-flight fracture risk could be. For muscle atrophy, exercise countermeasures are helpful, and people who spend months in space do fully recover their strength. It is unknown however, if 0.38 g on Mars will help reverse the muscle atrophy of the outbound flight in time for the return flight. This would become clear in lunar studies before a Mars mission.

Bone loss begins with weightlessness and continues throughout the mission. The excretion of calcium increases, as does the risk of kidney stones. The loss of calcium reflects skeletal remodeling concentrated in the weight-bearing regions. In the lower spine, hips, pelvis, and legs, bone loss will be 1.0 to 1.5 percent per month. Over six months, the astronauts would lose 6 to 7.5 percent of skeletal mass, but at 0.38 g, what would happen is anyone's guess. At best, the rate of bone loss would fall to near zero. On the return trip, however, another 6 to 7.5 percent would be lost, for a total of at least 12 to 15 percent. The bone experts at NASA do not seem to know what to do to prevent this yet, but drug therapy and artificial gravity are being taken seriously as countermeasures.

After the astronaut returns, the implications of this bone loss become clear when age and bone mineral density is plotted on a graph (figure 10.2). The graph compares the lifetime of an Earth woman with a woman that leaves at age fifty on a three-year trip to Mars. It indicates that the Mars woman

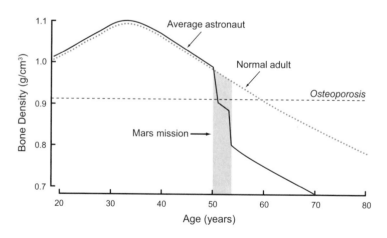

FIGURE 10.2. EXPECTED LOSS OF BONE MINERAL DENSITY ON A THREE-YEAR MISSION TO MARS FOR A TYPICAL FIFTY-YEAR-OLD FEMALE ASTRONAUT.
Source: Adapted from data in Clement (2003).

would develops osteoporosis about a decade earlier than if she stayed home. Clearly, the long-term risk of fractures must be factored into the decision to go.

ON BIOLOGICAL CLOCKS

Internal clocks set the physiological cycles in plants and animals and regulate many functions in biology. The study of biological clocks or the rhythms of living organisms is called chronobiology. The most famous clock, the twenty-four-hour *circadian rhythm*, is governed by external stimuli called *zeitgebers* (time givers). The zeitgeber synchronizes the internal clock with local time, and *daylight* is the most important cue.

The daily or diurnal biological clock in mammals is set by the eye. Light falling on the retina stimulates the optic nerve, which imparts a periodicity to cells in an area of the brain called the hypothalamus (Berson et al. 2002). In the hypothalamus, these pacemaker cells reside in an area called the su-prachiasmatic nucleus (SCN) and are synchronized into integrated rhythms by so-called *clock genes*. Clock genes are activated regularly, and their mes-senger RNA encodes for proteins that coordinate the electrical oscillations of hundreds of neurons (Yamaguchi et al. 2003).

Clocks also exist in other mammalian cells, including those of the car-diovascular system (Young 2006). These clocks are not fully understood, but they do regulate blood pressure and other circulatory functions. A mutation that disrupts the heart clock in the mouse slows the heart rate and causes loss of normal diurnal variations in cardiac power (Bray et al. 2008). Such mice also exhibit changes in the expression patterns of hundreds of cardiac genes involved in cell regulation, protein synthesis, and metabolism.

The cardiac clock, it seems, is anticipating circadian differences in physical work and is thereby tuning its responses to its environment. The circadian rhythms adjust an organism's metabolic activity to its behavior at a particu-larly advantageous time of day. And these rhythms are required to learn to *anticipate* by incorporating local time and the behavior of the internal clock into learning and memory.

Diurnal as well as seasonal light cues are involved in the regulation of sleep-wake cycles, appetite, immune function, psychomotor performance, and endocrine function in people. Often, even normal gene expression and other basic biochemical processes show predictable diurnal variations

(Merrow et al. 2006). Loss of these cues affects fitness in space just as it does in unusual environments on Earth.

Most of us are fully cognizant of the effects of sleep deprivation, jet lag, and rotating work shifts. They affect our sense of well-being and our performance, but most importantly, the disruption of biorhythms contributes to diseases ranging from obesity to heart attacks. In astronauts on short missions, these rhythms are reasonably well maintained despite the twilight on the spacecraft. However, astronauts usually only sleep about six hours a night and eventually show signs of sleep deprivation. After about three months in space, the natural circadian rhythms tend to dampen out, but the reasons for this and their implications are unknown.

A question for Mars planners is how careful to be in maintaining normal daily rhythms on the mission. Once there, a Mars day is only about forty minutes longer than an Earth day, but Martian light levels are lower. Clock biologists don't know how this will affect the crew on a multiyear stint on the Red Planet, but by ignoring chronobiology, we may literally let the clock tick toward poor decisions, poor performance, disruptive behavior, or the emergence of disease. Our knowledge of biological rhythms is not yet sufficiently grounded to perceive the implications of this clearly.

PART 3
WHERE ARE WE GOING?

11. THE CASE FOR MARS

The title of this chapter echoes a series of symposia as well as Robert Zubrin's (1996) intriguing book. Zubrin and like-minded people are highly vocal about their view of Mars exploration: within a decade, they claim, current technology could carry us to Mars. Diametrically opposed are the folks who say that sending robots is cheaper and safer, especially since it is looking more likely that there has never been life on Mars. But here again we see the unnecessary dichotomy between man and machines.

Mars activists argue that the so-called Siren of the Moon has seduced NASA into ignoring the Red Planet. They see Mars as an interesting place and the Moon as dull, barren, inhospitable, and of little scientific or practical value. This faction includes Zubrin, whose credibility is bolstered by a professional background in aerospace engineering. Yet his so-called Mars Direct mission is no closer to reality than when it was first proposed in the 1990s. His analysis, although visionary, had some flaws, and even NASA's new first Mars mission, penciled in for 2035, won't happen unless our propulsion, power, and surface technologies are an iteration beyond the systems being prepared for the next twenty years. It is sobering to look over the two-page list of new technologies required for Mars outlined in the NASA 2011 HEFT report.

A journey to Mars is no voyage in a sailing ship on the high seas; it is more like Magellan circumnavigating the globe in one ship—with only an airless Antarctica from which to replenish his stock and make repairs to his ship. Magellan, five ships, and 270 men left Seville; three years later, one ship, the *Victoria*, and eighteen debilitated men returned—sans Magellan. Over 90 percent of the crew was lost, and *Victoria* barely made it. Such a "success" on Mars would shut NASA's exploration-class missions down for the rest of the century.

Robotic probes have been steadily accumulating more data about Mars. In doing so, Mars has begun to seem less exotic—and less hospitable—than we had hoped. The case for Mars has slowly lost steam as it has become clear that traveling millions of miles in deep space, landing on another large planet, and coming home again comes with higher risks and higher costs than we can handle right now—regardless of the payoff.

The idea of getting there and back quickly has much to recommend it because it reduces the periods of discrete risk for radiation, weightlessness, and life-support malfunctions. This means waiting until the travel time is *short enough*, but no one yet knows how short is enough, or how to shorten it. And really fast travel depends on advanced engine designs that are still on the NASA drawing boards. It is not clear how expensive or how successful these will be in the long run.

Chemical rocket engines are hampered by a low efficiency, or a low *specific impulse*. Specific impulse is measured by how long a pound of fuel will produce a pound of thrust. For instance, the Space Shuttle's main engine had a specific impulse of about 450 seconds. With enough fuel to get to Mars, this engine would get us there in six months. In principle, advanced solar electric or nuclear electric drives could cut this travel time significantly. It may also be possible for the ship to accelerate and decelerate for enough of the trip to produce artificial gravity in the cabin, which would ameliorate some of the detrimental effects of weightlessness. The idea of generating gravity on interplanetary missions, which makes the spacecraft harder to control, is unpopular with aerospace engineers, but it may be necessary for medical reasons. The problem with bone loss has already been emphasized, but there is also evidence that the "puffy-face bird-leg syndrome" is not as benign as I earlier led you to believe. The facial swelling from the cephalic fluid shift in astronauts caused by exposure to microgravity is regularly accompanied by swelling of the optic discs, called *papilledema*. The optic disc is the round area at the back of the eye where the optic nerve, which carries visual information to the brain, enters the retina. Since the skull is closed, moderate increases in the pressure inside it can cause the optic nerves to swell, affecting the vision. Indeed, about a third of the astronauts on short missions and two-thirds on long missions actually experience decreases in visual acuity that may persist for years after the flight (Mader et al. 2011).

It is thought that swelling of the face and neck slightly obstructs the drainage of venous blood from the brain, leading to mild but prolonged increases in intracranial pressure. In this setting, bulging of the optic nerve or flattening of the eyeball may permanently and progressively affect the astronaut's eyesight. If so, you can expect to see partial gravity on manned spacecraft to Mars and beyond.

The idea of a premission to one of Mars' moons, Deimos or Phobos, has also become popular, and NASA first sponsored a workshop on this plan in 2007. There is a case for using the more distant Deimos, which has a nearly "geocentric" orbit, as an observation post. Its small size (less than ten miles across) has the advantages and disadvantages of working in microgravity, and its low density suggests an abundance of water ice or dry ice. Phobos is the larger and closer of the two, but it is still tiny (28 km in diameter). It revolves around Mars every 7.5 hours and is a mere 5,800 miles above Mars, and the views would surely be breathtaking. There would be much to learn about Phobos, whose origin as a moon is quite perplexing, but imagine going the entire way to Mars without actually landing on it! My friend "Q" has likened this to kissing his sister.

I have tiptoed around the issue of getting the crew off the planet for their return journey because it is a big problem for experienced engineers. The Apollo approach is attractive because it worked, except that the escape velocity for Mars is a daunting 11,250 mph (5.03 km/sec). Mars has 38 percent of the gravity of the Earth, and it retains a wafer-thin atmosphere. The planet's atmosphere does create drag, and some winged flight is theoretically possible. Again, the aerospace engineers are working on some options.

If history is any indication, the 2035 target date will slip, but the amount of slippage will depend on what we do with our new SLS. If NASA is not engaged on the Moon but busily chasing asteroids, we should not be surprised when China lands on the Moon, develops the necessary surface technology, and then takes it to Mars before we do. It is also conceivable that the first

Mars mission will be privately funded, but the approach most likely to succeed is the development of focused partnerships among the invested nations and the aerospace industry.

Distance aside, Mars has one big advantage over the Moon: large areas of water ice, particularly at the poles. Its other advantages are less critical, such as more O_2 in the soil, a twenty-four-hour day, an atmosphere, and better temperature stability. The amount of sunlight for power, although not up to the lunar level, is adequate.

As the complexity of the cosmic radiation issue has become clear, it has also made NASA scientists increasingly nervous. NASA is working on shielding strategies, but it has not yet decided on what to do. New lightweight materials, water, and magnetic shields generated by portable electromagnets are under investigation.

The magnetic shield is based on a strategy originally proposed for Apollo, and it might be done with a reasonably small electromagnet that could fit aboard a Mars spacecraft (Chown 2010). But before sending people to Mars on an electromagnetically shrouded ship, the idea would have to be tested outside the Van Allen belts with a working prototype, for instance in lunar orbit or on the Moon. If it works, another twenty years might see us on Mars, after upgrading the transportation technology currently in the works.

Portable magnetic shielding for a lunar habitat may be effective, although expensive, but locating a habitat on the Moon or Mars, where there are already minimagnetospheres, would be less expensive. Several such naturally protected areas are known on the Moon, probably the remnants of ancient cataclysmic magnetizing events on the lunar surface, and some are up to one hundred kilometers wide.

When Zubrin wrote *The Case for Mars* in 1996, he also understood the importance of establishing an advance habitat well before the mission leaves Earth. His rationale was fuel, life support, and safety, as well as cost. Despite the lead time, this part of his plan is compelling because getting off Mars is no picnic. NASA engineers are working along the same lines but are less enamored with setting up a refueling station on Mars for the return flight.

The Mars habitat would be stocked with O_2, water, food, and other resources for the crew for eighteen months. The crew would explore the region, maintain the habitat, and learn to deal with the human health issues. These issues epitomize the challenges of staying there.

THE WEATHER FORECAST

Martian climate and weather are also genuine concerns (Leovy 2001). Martian dust storms are formidable, sometimes global events that can greatly diminish the sunlight at the surface for long periods. Some of the most spectacular events originate in the southern hemisphere, in the Hellas Planitia, a massive impact basin covered in red dust. On the other hand, the atmosphere offers an alternative source of power: wind.

The pressure of the Martian atmosphere is less than 1 percent of that of Earth at sea level, about six mbar (4.57 mmHg or Torr). Like the Earth, barometric pressure falls with altitude, and at the summit of Olympus Mons, the pressure is less than one mbar. At the bottom of Valles Marineris, four miles deep, it may be ten mbar. The composition of the atmosphere, as measured by the Viking spacecraft in 1976, is 95.3% CO_2, 2.7% N_2, and 1.6% Ar. There are traces of O_2 (0.15%), carbon monoxide (0.07%), H_2O (0.03%), methane, and other gases.

Although the atmosphere is mostly CO_2, it is too thin to produce more than a weak greenhouse effect. The average temperature on the surface of Mars is only 5°C higher than it would be with no atmosphere at all, and the planet is as cold as Antarctica. For the human explorer on Mars, a continuous source of power will be just as critical as it at the South Pole.

Just like the Earth, the temperature on Mars varies with altitude, latitude, and season (Hess et al. 1979). Martian temperatures resemble our terrestrial deserts but with greater diurnal and seasonal variations because of the low barometric pressure and the absence of the moderating effects of our oceans. Atmospheric pressure on Mars increases as much as 25 percent in the summer because CO_2 sublimates at the poles (Malin et al. 2001). Moreover, the planet's axial tilt and orbital eccentricity, major determinants of the seasons, are greater than ours. Mars's tilt is about 25° compared with our 23.5°, and the distance between Mars and the Sun can vary by as much as twenty-six million miles, an orbit five times as eccentric as Earth's.

The atmospheric temperature on Mars changes daily by about 60°C compared with 30°C on Earth. At the poles, the daily surface temperature fluctuation is 90°C, while here, even in Antarctica, it is less than 50°C. NASA's exploration rovers confirmed the large seasonal and diurnal temperature variations at two locations, including near the equator (figure 11.1).

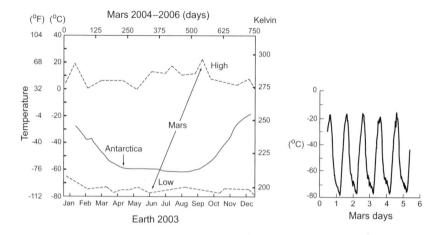

FIGURE 11.1. ATMOSPHERIC TEMPERATURE ON THE SURFACE OF MARS MEASURED AT 2°S LATITUDE.
Temperatures for a Martian year (687 Earth days) are on the top axis of the large graph. Air temperatures on the Antarctic plateau on Earth at 80°S latitude are plotted for 2003 on the bottom axis for comparison. The small graph at the right shows typical daily temperature fluctuations one meter above the surface for a six-day period near the equator.
Source: Courtesy JPL, Pasadena, Calif.

The search of the rugged terrain for future robotic and human landing sites and for niches for extremophiles has detected ancient hydrologic activity on Mars's surface. Mars was once covered by fairly deep seas, as evidenced by the extensive sulfate deposits and the famous "blueberry fields"—hematite spherules—discovered by the Mars rovers. These concretions represent gradual precipitation of the mineral out of aqueous solution and, in some places they cover thousands of hectares (Moore 2004).

A lot of attention has been given to Earthlike topography in the search for past riverbeds, lake sites, and springs. However, even crater rims and exhumed joints and faults show evidence that subterranean liquid once circulated there (Okubo and McEwen 2007). Layered deposits show signs of bleaching and cement formation probably reflecting the periodic flow of reducing and oxidizing fluids, similar to those found on Earth. These sites may have persistent subsurface water that would be useful to human explorers for practical and scientific reasons.

Yet most of the water on Mars is contained in frozen deposits at the poles, layered beneath caps of dust and dry ice—most prominently at the southern pole, which covers an area larger than Texas (Bandfield 2007). In 2007, these deposits were mapped by radar on the *Mars Express Orbiter*, and the echoes

208

suggested that the intervening layers contain approximately 90 percent water-ice perhaps up to 2.3 miles (3.7 km) thick. If this is correct, the southern pole contains enough ice to cover the planet's surface with water thirty-six feet (11 m) deep. This ice, however, is extremely cold—winter temperatures at the Martian poles fall below that which causes CO_2 in the atmosphere to precipitate as dry-ice snow.

The Martian water and dry-ice cycles, first detected many years ago, were confirmed by the *Phoenix* lander before it was frozen during the Martian winter (Hand 2008). *Phoenix* had detected a moderately alkaline soil (pH 7.7) highly enriched in salts of perchlorate (ClO_4^-), including magnesium and calcium perchlorate (Hecht et al. 2009). These salts are highly desiccating, but the soil neither prohibits nor promotes life as we know it. There are a few types of anaerobic bacteria capable of harvesting energy from the perchlorate molecule, but perhaps most importantly, the *Phoenix* data confirmed abundant O_2 in the soil of Mars.

Phoenix also found evidence of water in the Martian microenvironment, mostly in the form of hydrous phases or minerals (Smith et al. 2009). It is not clear how much water is absorbed or exists as ice cement in soil interstices during local changes in weather. Indeed, *Phoenix* found that high cirrus clouds (~4 km altitude) and ice precipitation (snow) allow water exchange between the atmosphere and the ground (Whiteway et al. 2009).

The water cycle on Mars is unusual compared with that of Earth. The atmospheric pressure on Mars is so low that liquid water exists only in the range of temperatures between 0° and 10°C. To understand this, we need a bit of chemistry and some meteorological information. Because of Mars's CO_2 atmosphere and eccentric orbital mechanics, there are large variations in the amounts of dry ice and water at the poles throughout the year (Byrne and Ingersoll 2003). Like the Arctic, the Martian poles recede in summer and expand in winter. Also, CO_2 and water molecules go through phases of gas, liquid, and solid, defined by the phase diagram and triple point, where the three phases meet. These phase diagrams are shown in figure 11.2.

On Earth at sea level, liquid water exists in the range of 0 to 100° C. The triple point, where liquid water, ice, and water vapor converge, is at 0.01°C and 6.1 mbar (0.06 atmospheres). This is close to the atmospheric pressure and surface temperature of Mars, but since the surface is usually colder than 0°C, water exists primarily as ice or as water vapor, and ice sublimates into the atmosphere. This is why water clouds appear in the Martian atmosphere and ice is found on the ground but liquid water is hard to find.

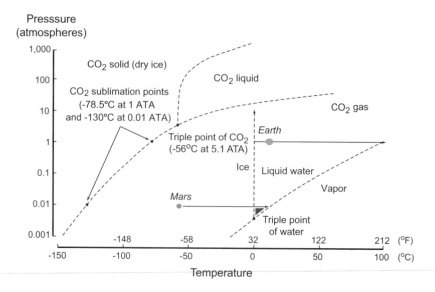

FIGURE 11.2. THE CHEMICAL PHASE DIAGRAMS FOR CARBON DIOXIDE (CO_2) AND WATER (H_2O) AT THE TEMPERATURES AND ATMOSPHERIC PRESSURES OF MARS AND EARTH.

The Y-axis is a log scale. The boundaries for the three major phases of each ice, liquid, or vapor are indicated by dashed lines. The average temperatures and pressures on the two planets are the round spots. Water can exist naturally in all three phases on Earth but mainly as ice or water vapor on Mars because of the low atmospheric pressure. Water on Mars behaves like dry ice, and it boils at 10° C (50° F). Only in the narrow range between 0 and 10° C (see solid triangle) is water liquid on Mars.

Water on Mars thus behaves like dry ice (CO_2) on Earth. CO_2 on Mars, because of the lower atmospheric pressure, sublimates (ice to vapor) at $-130°C$, compared with $-78.5°C$ on Earth. The dry ice/water ice mixture at the Martian poles is thus colder than South Pole ice, and it would require a lot more heat to melt it to recover the water. This is one reason why a warmer spot than the poles is desirable for a Mars base.

Martian soils have a high salt content, which depresses the freezing point of water perhaps by 60°C or more, which could allow liquid brines to exist in some areas. In the mid-latitudes in the Hellas region, there are deposits with radar properties consistent with massive water-ice structures that may be debris-covered glaciers. Mid-latitude glaciers may have formed during previous climate cycles that were more conducive to glaciation and may represent significant amounts of nonpolar water ice on Mars (Holt et al. 2008).

Mars's greenhouse effect, unlike the Earth's, is influenced more by its red dust than by CO_2. At the end of the twentieth century, the brightness or albedo of Mars decreased, and the darkened southern highlands began to absorb more sunlight, causing Mars to warm slowly. A simulation of the planet's climate suggested that the mean atmospheric temperature increased by 0.65°C between 1976 and 2000 (Fenton et al. 2007), although to just −77°C (196 K). The warmer summer temperatures at high southern latitudes could be helping the steady retreat of the south polar ice escarpment, which has been underway since at least 2000.

The redistribution of dust by the wind causes large areas of Mars to darken or brighten. The exposure of dark bedrock by the movement of brighter dust into small depressions contributes to the surface darkening. Dark surfaces enhance the processes that raise dust and lead to further darkening, thus sustaining the cycle of warming.

Some Mars experts think that one thousand years from now the mean atmospheric temperature might naturally reach −50°C, causing polar CO_2 to sublimate and triggering further warming. This would leave mostly water ice at the poles. On the other hand, warming would increase wind speed and perhaps cause reflective dust to remain airborne longer, counteracting the warming trend. In other words, Mars, like Earth, is undergoing climate change, but planetary scientists do not know whether or not it will become a more temperate place.

Based on the distance and time in space, I have emphasized missions that put astronauts on the planet for eighteen months. The long length of stay and narrow departure windows are dictated by Earth-Mars orbital mechanics. The crew cannot simply leave whenever they want because they must conserve fuel and time in space. Eventually, O_2, water, and food must be produced on Mars, at temperatures comparable to Antarctica's. The longer the crew remains on Mars, the longer it must depend on vital surface technologies and the more likely it is that someone will have a serious injury or illness requiring, for example, a major operation.

SURFACE TIME

The conjunction mission minimizes time in space and maximizes time on the surface of Mars. A *fast transit* means the astronauts spend less time in microgravity and receive less cosmic radiation. The cumulative dose of radiation is reduced because more time is spent under the thin but protective umbrella of Mars's atmosphere. Despite its greater distance from the Sun, radiation is more intense on Mars than on Earth because of Mars's weak magnetic field. Also, high-energy cosmic particles easily pass through the thin atmosphere, heavily exposing Mars's surface to GCR and SPE. This means more secondary products are generated in the atmosphere and by interactions with minerals on the surface.

The atmosphere does deflect and attenuate some charged particles and provides better radiation protection than the Moon (Rapp 2006). The bad news is that many secondary neutrons are produced by cosmic rays that hit the ground. For low-energy GCR, more neutrons are propagated backward from the ground than are produced in the atmosphere. During a massive SPE, even high-energy neutrons may propagate as much radiation from the ground as from the atmosphere. Mars's radiation levels therefore vary greatly with the cosmic ray flux, atmospheric conditions, and the type of surface material. This makes the radiation dose on the surface difficult to estimate, and it varies from region to region. The average astronaut in a typical spacesuit at the equator would get roughly one hundred times more radiation than on Earth.

Although the Martian atmosphere, just like the Moon, contains virtually no O_2, there is plenty of O_2 locked in the soil (Yen et al. 2005). The soil is rich in iron oxides and sulfur dioxide, especially hematite (Fe_2O_3) and olivine (Mg, Fe_2SiO_4). Other metal oxides, such as aluminum, calcium, and titanium, are abundant too. As with lunar soil, O_2 can be released fairly simply by heating in the presence of a catalyst. Developing a simple technology to harvest and store O_2 in quantity from Moon resources is crucial to the exploration of Mars.

The use of greenhouses to harvest O_2 on Mars requires the establishment of a large biomass, so green plants at first would decorate the Martian dinner plate. The creation of greenhouse and soil processing systems that extract O_2 and add nitrogen for plant growth could provide two sources of O_2, and a safety margin. The development of surface soil science and a plant-growth facility (SPGF) are the next steps in sustainable bioregenerative life support, and these would actually be easier on Mars than on the Moon because of the former's relative abundance of CO_2 and water. There are also technologies for making artificial soils from lunar and Martian regolith minerals.

SPFG construction is a challenge to materials science because the structure would need translucence, great strength, and durability, taking into account mass, volume, power, water, and maintenance requirements. Greenhouse mass can be reduced by inflating flexible or semirigid structures with CO_2 to a slightly greater pressure than the Armstrong line (plants will desiccate in a near-vacuum, too). A bit of O_2 can be bled in until photosynthesis generates enough O_2 to support both plants and people (Hublitz et al. 2004). Technologies already exist for the fabrication of transparent, inflatable domes that use both natural and artificial lighting to grow plants.

A fairly straightforward approach to optimal SPGF horticulture has already been worked out. The optimal output of edible biomass, called Q_{max}, is related to the ratio of crop yield to energy cost. The implication is that food production is most efficient when the crop volume and the energy cost are low and the growth period is short. However, artificial systems are still inefficient, which means they require more energy than natural systems.

If you are interested in edible biomass, the expression $M \times EBI^2 / (V \times E \times T)$ gives you the Q_{max}, where M is the total crop harvest (dry biomass) and EBI^2 is the edible biomass (index) squared. These variables are multiplied and then divided by the product of V (the volume of crop), E (the energy for crop growth), and T (the growth period) (Berkovich et al. 2004).

The transfer of lunar surface technologies to Mars should be straightforward. Transferable technologies include habitat structures, tunneling and mining equipment, ground transportation, and spacesuit technology (Yamashita et al. 2006). Regrettably, the atmospheric pressure on Mars is too low for humans to do without pressurized habitats and pressure suits, and breaches of habitat integrity would cause hypoxia, the bends, and ebullism, just as they would on the Moon. Similarly, Mars habitats, vehicles, and spacesuits will intrinsically lose gases that must be compensated for over time.

The use of underground tunneling or natural underground structures on Mars offers the same advantages as on the Moon for protection from radiation and the retention of heat and atmospheres. NASA has funded the evaluation of this concept for some time, and the approach could be combined with situating near magnetic hot spots that limit the average amount of cosmic radiation hitting the ground.

In short, the initial exploration of Mars will require faster transportation, effective shielding and radiation countermeasures, experienced crew selection, a carefully chosen landing site, and sound surface and subsurface technologies. The use of localized magnetic hot spots on the planet could provide significant shelter from cosmic radiation and would be especially important for the selection of a site for a permanent Mars base. Solving these problems by the end of the century might increase the time humans could spend safely on Mars to as much as a decade.

SUITS AND STRUCTURES

The Advanced Development Office at NASA's Johnson Space Center has been working on novel concepts for planetary habitats for years. The habitats and spacesuits for Mars may evolve to incorporate different materials, thicknesses, strengths, and thermal properties than used on the Moon, but strong, flexible lightweight materials that prove serviceable on the Moon should transfer well to Mars. In addition to being lightweight and durable, the structures must be easy to install and operate, spacious, and inexpensive to maintain. The preferred concept, a preintegrated, semi-hard-shell module set up on the lunar surface before the mission crew arrives, would work on Mars too. Powered landing systems like the one developed for the *Curiosity* rover could be perfected for habitats. Alternatively, soft, self-assembling, prefabricated structures could be bounced onto the planet's surface and deployed or inflated automatically.

Ultimately, structures of indigenous Martian or composite Mars and Earth materials will be needed, and the robotic technology for subsurface tunnel boring and regolith processing worked out for the Moon would be transferable to Mars with a little modification. Future habitats have also been envisioned that would assume certain biological features. These may include intelligent self-diagnostic and corrective or repair capabilities or incorporate self-renewing biological materials into the structures.

The habitat is the centerpiece of Mars exploration, yet the importance of ground transportation and spacesuits cannot be overlooked. The procedures worked out for habitat entrance and egress and protection from lunar dust will be invaluable on Mars. Mars vehicles and spacesuits must have extended capabilities, and for an astronaut to venture any distance from base, both must be improved significantly (Harris 2001). A spacesuit has more constraints than a passenger vehicle, although hybrids—powered suit systems—similar to the extravehicular mobility unit (EMU) used in LEO are a good bet for Mars.

Today's EVA suits are limited not only by mobility but by mass and size. A Mars suit will be smaller, lighter, more flexible, and have a lower center of gravity than a traditional suit while still offering durable protection against the environment, particularly dust. This also applies to accoutrements such as inner garments, helmets, visors, and gloves.

A pressurized, puncture-resistant, self-contained Mars suit is critical because the atmospheric pressure on Mars is far too low to relax suit stringency. Unfortunately, notions of elastic body suits and pressurized O_2 masks are physiologically untenable. The Armstrong line of Earth's atmosphere is an order of magnitude higher pressure than the Martian atmosphere. Just like today's suits, Mars suits will be called on to eliminate CO_2, regulate temperature, provide O_2 and H_2O, and contain waste.

The notion of converting Mars into a planet more suitable to our biology is popular in science fiction because "terraforming" is deceptively simple if you overlook the absolute mass and temperature of the planet. Many ideas have been put forth, and theoretically, some are possible, but none are technologically and economically practicable. The atmosphere of Mars is in thermal and chemical equilibrium with the planet's surface, and over millions of years may have settled out at a minimum near the triple point of water (6.1 mbar). If so, it may be tough to move it off of its settling point without help from the Sun and geological time.

Mars's climate changes naturally because of its axial tilt. When the southern ice cap is exposed to the Sun, enough CO_2 may be released to nearly double the atmospheric pressure (Phillips et al. 2011). Martian soils also contain enough CO_2 bound as magnesite ($MgCO_3$) that releasing it could hypothetically increase the planet's atmospheric pressure to more than two bars (1 bar = 0.987 terrestrial atmosphere). The greenhouse effect would push the temperature up, but a higher temperature and pressure would increase the volume of surface water into which CO_2 could dissolve, which might oppose a favorable shift in the atmospheric equilibrium. Some O_2 would also have to be released into the sky from soil and later from green plants.

Converting the atmosphere from cirrus to nimbus would be a vast undertaking, but how about increasing Mars's pressure just enough to get rid of our pressure suits? In other words, what would it take to get Mars's atmosphere below the Armstrong line? The line is 62.8 mbar, and the average pressure on Mars is 6.1 mbar. Even the tenfold increase in pressure, however, would not provide a breathable atmosphere, even with pure O_2 and a pressure-demand mask and regulator.

An acclimatized Sherpa on Mt. Everest, where the total pressure is 336 mbar, has an inspired PO_2 without extra O_2 of about 70 mbar, or 11.5 times higher than the total pressure on Mars. An Everest mountaineer with extra

O_2 has an inspired PO_2 of about 110 mbar, or eighteen times greater than the total pressure on Mars. A military pilot's pressure-demand O_2 mask (designed for 43,000 feet) can be used in an emergency at up to 50,000 feet (117 mbar). Thus, depending on the level of acclimatization and the type of pressure-breathing system you choose, the pressure on Mars would have to be increased 12.5-fold with O_2 to become directly breathable or eighteen- to twenty-fold without it in order to breathe with bottled O_2 and a pressure-demand mask and regulator. Given Mars's diameter, a breathable atmosphere a little over two kilometers thick would fill a volume of three hundred million cubic kilometers (seventy-two million cubic miles) in round numbers.

ONE-WAY TRIPS

Mars is far away, cold, and has a minimal atmosphere and a higher radiation profile than once thought. Astronauts will take great risks getting there and staying there. After eighteen months on the planet, they would enter their return vehicle, launch it, escape the planet's gravity, and spend six more months in space coming home. It is difficult to predict when the technology to minimize these risks will become available, and there may not be enough information to make an educated guess until well into a new phase of human exploration. Alternatively, one might throw caution to the wind and send a few old volunteers on a one-way mission to establish a habitat on the planet. This idea has been floating around for some time, and it is not as outrageous as it might seem at first blush. It is more likely, however, to be put into action by private rather than government financing.

Proposals began appearing in print about 2006 for the one-way, one-person option, which was proposed by the former NASA engineer James C. McLane III, and the "Mars to Stay" mission was pushed by the former Apollo astronaut Buzz Aldrin. These bootstrap ideas are designed to reduce the cost and boost the speed of settlement of Mars. Traditionally, explorers have risked their lives for lower odds, and driven by opportunity, migrants often leave home with no intention of returning. As the days of staking a claim for church or country are long past, a one-way Mars voyager would obviously have an unconventional mindset and be driven by unconventional priorities. Only time will tell, but humanity's first landing on another planet could be chronicled by the smartphone.

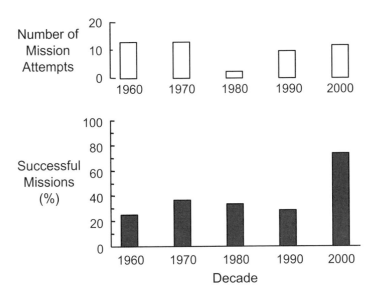

FIGURE 11.3. THE ROBOTIC EXPLORATION OF MARS SINCE 1960.
The top graph indicates the total number of probes launched from Earth toward Mars in each decade. The bottom graph is the percentage of missions that planners rated as successful. The percentage of successful missions has been steadily increasing, but the missions have all been one way. No Mars return missions have been undertaken except the 2011 Russian *Phobos-Grunt* mission, which failed. The 2008 soft landing of the *Phoenix* lander was the first in thirty-two years and the third in history. Mars *Curiosity* is not included.

In thinking this over, it is instructive to review how unmanned Mars probes have fared in the past (figure 11.3). The history dating to 1960 shows steady improvement, but round-trip missions to Mars still have low odds of success. Since 2000, one-way lander missions have succeeded about 83 percent of the time, but this will not do for a mission involving people. If the one-way success rate is 83 percent, the round trip success rate, everything else being equal, would be roughly 69 percent. Recall that the shuttle, over thirty years, had a 98.5 percent success rate. The one-way lander information does not take into account the time on the planet or the departure from the surface. However, if the one-way success rate could be raised to 98.5 percent, there undoubtedly would be a line of volunteers.

In a word, I would stress three key ideas about Mars. First, time and distance make any Mars mission dangerous, and even a modest risk of failure is unacceptable because it would expose the entire manned space program to being mothballed. The loss of a Mars crew would put an end to exploration-class missions for a long time. Thus the first attempt must have a high probability of success.

Second, paring these risks down requires time, money, and experience. The ability to reduce risk is closely connected to a clear understanding of the environment, rapid travel times, reliable life support and radiation shielding, and thorough contingency planning. Proper timing based on proven equipment is implicit, including prearranged habitats and surface technology.

Finally, a successful first mission and important discoveries made during it will govern the level of interest in returning to Mars. Mars is not a conquest, and a sound scientific agenda is essential. The alternative of sending a few space cowboys on a one-way trip is not completely out of the question for a private venture, or perhaps for China, where failing to "volunteer" has inimical undertones. However, the most sensible approach entails the formation of partnerships among the committed parties. In any case, Mars planners will know to select a region with the best mission characteristics and the best possibilities for groundbreaking science.

Although the dream of stepping onto the Red Planet to discover the first extraterrestrial microbe or bug is fading, other new discoveries may be just as staggering. The planet's resources are massive and could offer untold benefits to humanity given the wherewithal to handle the discovery issues as a unified, advanced civilization. Imagine your great-great-great-granddaughter Mary Esther one day telling her daughter: "Did you know the yttrium and terbium in bot-mentor's lemon-lime sun wings came from the Xenotime Mine in the Valles Marineris?"

12. BIG PLANETS, DWARF PLANETS, AND SMALL BODIES

We are approaching the point in our technological development where large-scale explorations of the Moon and Mars are becoming feasible twenty-first-century objectives. The outer system, however, is six times larger than the inner system, and beyond Mars reside four giants, hundreds of moons, and a tapestry of dwarf planets and small solar-system bodies, a few of which could be intriguing places to explore. We would need a special reason to send people to them, particularly since an entire other world, Mars, could preoccupy us for the next five hundred years.

For better or for worse, astronomers decided finally to define what a *planet* was in 2006, and we again live in a system of eight planets. A planet was previously an object in between a moon and a star, but the discovery of Pluto-sized objects in the Kuiper belt began to cause trouble. Historically, "small" objects like Pluto and Ceres have fallen into and out of planetary status for centuries, although astronomers knew that planets were really big. The outer planets were also distinguishable not just by their overstuffed size but by their gaseous natures.

In 2006, at the Twenty-sixth General Assembly in Prague, the International Astronomical Union (IAU) made up a new class of celestial object

TABLE 12.1. IAU 2006 DEFINITION OF A PLANET

CATEGORY	DEFINITION	MEMBERS
Planet	A celestial body (a) in orbit around the Sun and (b) of sufficient mass for its self-gravity to make it assume a nearly round shape and that (c) has cleared the neighborhood around its orbit	Mercury, Venus, Earth, Mars, Jupiter, Saturn, Uranus, Neptune
Dwarf planet	A celestial body (a) in orbit around the Sun and (b) of sufficient mass for it to assume a nearly round shape but that (c) has not cleared the neighborhood around its orbit and (d) is not a satellite	Ceres, Pluto, Eris, Makemake, Haumea
Small solar-system bodies	All other objects orbiting the Sun except satellites	Asteroids, comets, most trans-Neptunian objects (TNOs)

called "dwarf planet" and relegated Pluto to it. The IAU divided the Solar System's celestial bodies into three categories: *planets*, *dwarf planets*, and *small solar-system bodies* (table 12.1). In a nod to Pluto fans, dwarf planets that orbit beyond Neptune were named "plutoids."

As a strictly amateur astronomer, I gravitate toward the parsimonious view of the astronomer Mike Brown, the discoverer of Eris and other distant things (Brown 2010). According to Brown, we have learned a lot about the Solar System simply by watching how the big stuff behaves: a yellow star, four rocky inner planets, an asteroid belt, four gas giant outer planets, and a cloud of diverse objects, the Kuiper belt. This makes a lot of sense in view of the large number of other planetary systems that have already been discovered.

The inner and outer systems are separated by a void between Mars and Jupiter populated by thousands of remnants of the ancient Solar System, the main asteroid belt. Mars, comparable to Antarctica, is balmy compared with the objects here. The lowest recorded temperature on the Antarctic plateau is $-89.5°C$ ($-128.5°F$), or 183.5 K. Ceres is 167 K, Jupiter's moon Callisto is 120 K, and Saturn's moon Titan is 90 K. Uranus's Shakespearian twins, Oberon and Titania, are ~60 K. Neptune's Triton, the coldest moon of all, is 35 K. Recall that liquid nitrogen's temperature is 77 K; these are cryogenic conditions.

THE TRUTH ABOUT ASTEROIDS

Since the other side of the asteroid belt is so prohibitively cold—and because NASA has a mandate to visit an asteroid—let's think asteroids for a minute. There are a huge number of them around, and it is important to understand the scientific implications of such missions.

NASA brainstorming about choosing, intercepting, and sampling an asteroid has led it to design tethers, jetpacks, bungee cords, and cargo nets. Near-Earth asteroids move fast and have almost no gravity. Our would-be explorers would fly for weeks or months in a mother ship, drop from a shuttle-craft, hover just above the surface of the asteroid, and then take samples to bring home. The mother ship's engines, navigation, and shielding would be magnificent, but the rest of the technology that requires development will be expensive, highly specialized, and almost exclusively for use with asteroids.

The selection of an asteroid and the timing of the mission are contentious mainly because every other interesting place in the Solar System has significant gravity and requires vital surface technologies. I had to smile when, with a perfectly straight face, NASA's Near Earth Object program manager Donald Yeomans justified this mission to the press: "Every hundred million years or so an asteroid six miles wide—the type that killed off the dinosaurs—smacks Earth."

In 2011, the asteroid 2055 YU55, a rock the size of an aircraft carrier, did come within 202,000 miles of Earth, about three-quarters of the way to the Moon. This was the closest approach to Earth of an object of that size in thirty years. Had YU55 hit us, it could have created a four-thousand-megaton blast on land or a seventy-foot tsunami in the ocean. But are you so worried about such a cataclysm that you think it's worthwhile to rappel onto such a rock yourself? Why not just send a probe of some kind?

NASA's Near-Earth Object Program Office at the Jet Propulsion Laboratory (JPL) in Pasadena was established to detect, track, and characterize potentially hazardous asteroids and comets near Earth; hence the NEO acronym. Once found, an NEO is tracked for a while, and its orbital path is predicted. Computer models are then used to calculate the probability of it hitting Earth within next few years. The NEO Program Office website offers many interesting facts, which are worth a look, but the implications are left mainly to the imagination.

Every thousand years or so, a meteoroid the size of an aircraft carrier does hit us and causes impact damage to a large area, such as the Tunguska air burst in Siberia in 1908. The early detection of such a threat is in our best interest, and the options for deflecting such objects have been studied in some detail. NASA experts mention missiles and thermonuclear devices as "relatively mature options," but there is more than a little disagreement here. There is no disagreement, however, about human visitors having no role in deflecting asteroids.

A small subset of large asteroids passing close to Earth several years before the impact might be amenable to a nudge by a robotic gravity tractor placed near the asteroid. The presence of the tractor would produce a tiny change in the velocity of the asteroid that would be amplified by the gravitational effects of its flyby, causing it to miss us the next time its orbit takes it around.

The NEO office claims to have accounted for 60–90 percent of all civilization-wrecking objects, mostly asteroids, but only about 1,300 potentially hazardous asteroids have actually be catalogued so far. These objects are at least five hundred feet in diameter and pass within 0.05 AU (about 4.65 million miles) of Earth. However, the number of NEOs increases sharply as their size decreases. We are more likely to be hit by something too small and moving too fast to be detected in time to knock it off course. People in the path of the object, like those living along the track of a hurricane, would have to be warned and evacuated beforehand.

Recall, too, that the Earth is 70 percent water and that we inhabit less than 1 percent of its surface, mostly along coastlines. In the worst case, the next big meteoroid event would resemble the 2004 Indonesian earthquake and tsunami that killed 230,000 people. If NEO control officers stay alert, we should have time to move people out of the way and prevent mass casualties. On the whole, it seems more sensible to worry about a terrorist with a suitcase bomb or a deadly global pandemic than about being hit by a massive space rock.[1]

The main asteroid belt does not contain much mass because of the outward migration of the giant planets in the early Solar System. Many asteroids were ejected inward and hit the inner planets during the late heavy bombardment 3.9 billion years ago. The evidence is found in the many craters on the Moon dating to a relatively short time span 600 million years after the formation of the Solar System (Walsh 2009). A repeat volley of asteroids at this point would require a major shakeup of the outer system.

The composition of the remaining asteroids is related to their distance from the Sun. Astronomers originally typed asteroids by chemical composition, using a classification called CSM. Nearly 80 percent are C-type, especially toward the belt's outer edge, similar to carbonaceous chondrites. Most of the rest, found at a distance of 2 to 3 AU, are stony or S-types, made of nickel-iron and iron-magnesium silicates. A few are shiny metallic M-types, almost pure nickel-iron. There are overlapping forms and a dozen rarer types but all lack precious stones or metals, despite the latest pipe dreams of would-be asteroid miners.

OASIS CERES

Since no killer asteroid is threatening us right now and I can't tell NEOs apart, let's look at the interesting dwarf planet Ceres. Ceres may be useful if we ever need a stopover to help explorers cross the asteroid belt, but this is an "if" in capital letters. Regardless, relative to other round bodies in its vicinity, Ceres' characteristics are exceptional.

A trip to Ceres is 3.5 times lengthier than one to Mars. Ceres orbits the Sun every 4.6 years at roughly 2.76 AU, closer to Mars than to Jupiter. It is never closer to us than 160 million miles and always invisible to the eye, but it is six hundred miles across and contains one-third of the mass of the asteroid belt. Discovered by the Sicilian astronomer Giuseppe Piazzi in 1801, it was hailed as a new planet, but soon other tiny planets were found, and they all were renamed asteroids. Apart from its upgrade to a dwarf planet, Ceres attracted little attention until the Hubble was turned on it, and in 2005, something unusual was seen.

Ceres' shape is similar to Earth's; it too pulled itself into a sphere under its own gravity (which requires a diameter of about 250 miles or 400 km). It is one of four round objects in the asteroid belt; Pallas, Vesta, and Hygeia are the others (McCord et al. 2006). Planetary spheres are oblate spheroids—flattened at the poles and slightly fat at the equator. This shape represents the equilibrium between the inward force of gravity and the outward push of internal pressure. The equatorial bulge is caused by rotation. A planetary mass is optimally organized for the forces acting on it and is therefore *relaxed*. Ceres is the only relaxed body in the asteroid belt.

Ceres is not homogeneous; it is layered or differentiated (Thomas 2005). It has a rocky core and an enveloping water-ice mantle perhaps 100 km (62 miles) thick and covered by soot—a dirty snowball. Ceres' density is about 2.1 g/cm^3 and less than Vesta and Pallas, which are rocks. The density of water ice is about 0.94; rock is about 3.5, and most metals around 4.5 g/cm^3. Ceres is also more reflective than its neighbors, reflecting 11 percent of the sunlight that strikes it. This means Ceres is up to 27 percent water.

Where did all this water come from, and how has Ceres, with no atmosphere, held on to it for millions of years? The ice should have sublimated, but it did not, probably because of Ceres' solid crust of mineral clays, mostly carbonaceous chondrites and saponite from the stellar nebula that formed our Solar System.

In 2007, NASA's *Dawn* spacecraft set out to photograph, map, and measure the gravitational fields and density of Ceres and Vesta. Vesta is a quarter the mass of Ceres but closer to Earth; it has a geology like the four inner planets. Ceres is like the moons of the outer planets. Learning more about the differences between the two should tell us more about the transition between the inner and the outer Solar System. *Dawn* reached Vesta in 2011, and NASA cooked up a "*Vesta* Fiesta" for the kids before the craft moved on to Ceres, where it will arrive in 2015. *Dawn*'s novel xenon ion engines and its solar-powered instruments have been of great interest to engineers, but riding for more than eight years on a *Dawn*-like ship would kill you from boredom or radiation.

If you had to go there, you would find resources on Ceres, more fresh water than Antarctica has, and room to explore: it has roughly the surface area of Alaska plus Texas. It has no atmosphere, and the ship would land easily. It also has a low escape velocity, only 0.51 km/sec (~1,100 mph). Ceres gets enough sunlight for standard collectors to generate solar electricity, but your collector would need to be 7.7 times larger than one on the Moon to generate the same amount of power.

Ceres has a nine-hour day, and the Sun in the sky, one-third the usual diameter, would rise and set quickly. Ceres is cold, averaging −106°C (167 K), although it does warm up to −34°C (−23°F or 239 K) during the day. At sunset, the temperature plunges 70°C, and the morning (or evening) star, called Earth, would be a blue Mars-like orb. You would need a pressure suit, but, weighing less than five pounds in its 2.8 percent gravity, you could learn to leap four stories. This also means that bone-loss and muscle-atrophy countermeasures are needed.

In science fiction, if you don't mind meteoroids and cosmic radiation, Ceres is the place from which to mine asteroids. The notion of asteroid mining is so prevalent that some think it is inevitable, but the asteroids have a combined mass of only about 5 percent of the Moon, spread out over ten times the area of the inner Solar System. It would be impracticable to recover much of anything there. Ceres does have magnesium, iron, silicon, ammonia, nickel, sulfur, and O_2 but nothing for the prospector more precious than peridot.

TITAN AND THE GALILEANS

The trip from Earth to Ceres took the *Dawn* spacecraft more than eight years, and Ceres is less than halfway to Jupiter. If *Dawn* continued on to Jupiter, it would add almost eleven years to the trip. Chemical propulsion is too slow and inefficient for the outer Solar System. Newer electric rockets propelled by plasma generate less thrust, but the fuel goes ten times as far (Choueiri 2009). Making such technology faster is the work of future aerospace engineers, as is determining the routes of travel for which a certain power achieves the highest velocity.

High-efficiency routes may exist in a network of virtual "tubes" defined by the combined gravitational fields of all of the objects in the Solar System. These gravitational hubs, euphemistically called an interplanetary superhighway, define the system's gravitational topography, along whose contours future spacecraft might operate (Stewart 2006). At the birth of the Solar System, the inner planets accreted from solar nebula between the massive gravity wells of the Sun and Jupiter. Jupiter is an abortive brown dwarf star condensate consisting mainly of the two lightest and most abundant elements in the universe, hydrogen and helium, in a ratio similar to that of the Sun. It may still connect to the inner planets through gravitational zones that could be used to boost travel efficiency to its moons. If so, we might bypass Ceres entirely.

The four giants are nothing like the four rocky inner planets. Saturn, also made up mostly of hydrogen and helium, is a *gas giant* like Jupiter. Uranus and Neptune are *ice giants* made of heavier elements, including carbon, O_2, and N_2. The size of these planets is defined by the tops of their atmospheres, since they lack a clear solid surface. Near their centers, the gas becomes denser, liquefies, and then solidifies. The composition of the cores, except for Jupiter's, whose is of metallic hydrogen, are unknown. The massive gravity and lack of a surface prevent "landing," unless a gas-liquid interface allows for the deployment of a submarine.

Approach Jupiter at your own peril, not just for fear of ponderous gravity, lack of a surface, and violent winds, but because the planet's powerful magnetosphere generates intense radiation. Jupiter's magnetosphere is the second-largest structure in the Solar System, after the heliosphere. It trails out behind the planet and stretches almost to Saturn.

Years ago, it was found that the planet gives off twice as much heat as it receives from the Sun. Jupiter's center of liquid hydrogen is in a highly conductive state, and the planet spins furiously, every 9.9 hours, generating internal magnetism and heat. This creates Jupiter's magnetosphere, a magnetic field fifteen times stronger than the Earth's, which traps charged particles in toroid belts encircling the planet (figure 12.1).

Jupiter's big radiation belts have big implications too: they will kill you. One belt between the sky and Jupiter's ring is a field of charged particles streaming from Io, while a second belt of neutral particles probably originates from Europa. The inner belt is ten times the strength of the Van Allen belts and generates massive auroras at the poles. The Cassini spacecraft determined that these radiation belts were 2.5 times more powerful than first thought. So if you want to see Jupiter, you must pick from its menu of sixty-three or more satellites.

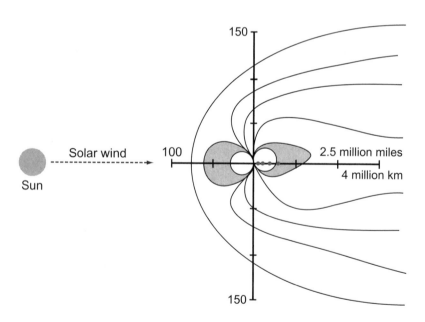

FIGURE 12.1. JUPITER'S MASSIVE MAGNETOSPHERE.
The axes are scaled in multiples of Jupiter's radius (71,434 km), with Jupiter located invisibly at the axis crossing. Cross-sections of the main radiation belts are shown by the shaded areas. The approximate positions of the four Galilean moons are shown as dark dots.

The four Galileans, Io, Europa, Callisto, and Ganymede, are all interesting, but only Ganymede and Callisto could be visited by people, because Io and Europa are too close to Jupiter. All are tidally locked, the same side always facing Jupiter, just as one side of our Moon faces us. The Galileans accreted from an ancient water-ice-rich skirt gathered around the new planet. Other ices are probably not present because the temperatures are a bit high.

The water content of the Galilean moons varies with distance from Jupiter, possibly an effect of the temperature gradient arising from the heat of the planet. Their densities fall progressively, indicating increasing amounts of water. Io's density is 3.5 g/cm^3, rock and metal, while Europa is 3 g/cm^3, silicates and water. Ganymede and Callisto are just under 2 g/cm^3, half water.

Io is frighteningly close to Jupiter (1.3 times farther than our moon is to us); it is a dry hellhole whirling around the planet every 1.75 days. In a spacesuit, you would receive a lethal radiation dose in fifteen minutes; you might as well stand on the melted Fukushima reactor in your shorts. Io is surprisingly cold but has local hot spots caused by tidal interactions with Jupiter, Europa, and Ganymede. Io's interior flexes, generating heat and volcanism and giving the moon a thin sulfur dioxide atmosphere.

Europa is next, the smallest of the Galileans, 90 percent the size of Selene and with a bit less gravity (13 percent of Earth's). It orbits at some 416,900 miles (670,900 km), in a nearly perfect circle, every 3.5 days. A darling of science fiction, Europa has attained superstar status because of its apparent subsurface liquid ocean, which engenders speculations of life. This is an astonishing conjecture, given how cold it is and what little knowledge we have about the origins of life on our own planet, which has a far warmer history.

Nevertheless, a pioneer on the origins of life, the chemist Stanley L. Miller (1930–2007) at the University of California, who earned fame with Harold C. Urey, proposed that life could originate in the cold. Miller and two colleagues identified a variety of prebiotic building blocks of life, amino acids and RNA bases, in a dilute cyanide solution frozen away at −78°C for twenty-seven years. They felt these compounds could have been produced on the primordial Earth by a *eutectic mixture* of ice, ammonia, and hydrogen cyanide (HCN) (Miyakawa et al. 2007).

A eutectic mixture freezes at a lower temperature than any of its constituents. A good example is seawater, which is 0.35 percent salt. As a result, our oceans freeze at −2° C (28.4°F) instead of 0°C. The ice crystals are basically pure water, which means that microscopic pockets of liquid are trapped at a high salt content, allowing certain chemicals to polymerize in the cold.

Miller's notion was criticized for three reasons. First, as temperature falls, chemical reactivity falls; second, cyanide was rare in our primitive oceans; and third, polymers trapped in ice crystals probably won't generate life. The idea of a cold start for life on comets or icy moons has especially little to recommend it from biologists who favor warmth.

Europa too is cold; it averages $-170°C$ (103 K), and although it has a weak magnetic field and a trivial O_2 atmosphere, Jupiter's radiation belts would kill you in a day or two unless your ship punched quickly through the ice to find shelter. Cold ice is hard ice, and drilling through it requires high torque and power. Europa's crust is five times thicker than South Pole ice, but you would need only to excavate one hundred feet or so to escape the radiation. Nonetheless, these conditions may hide Europa's secrets for a very long time.

Europa, like a terrestrial planet, is an onion, and its density and topography suggest that the ocean sits 15 km (~9.5 miles) below the crust. The crust floats, and the high pressure causes the water to freeze at a lower temperature. The ocean could be stabilized by tidal heating and a high salt content. If so, Europa has the mightiest ocean in the Solar System (62 miles or 100 km deep). In Antarctica, deep ice covers Lake Vostok and other fresh-water lakes that could contain ancient microbial life, but that life could not have evolved there. Moreover, the Russians spent more than twenty years drilling through a mere 2.3 miles of ice to get a sample of Lake Vostok's water in 2012. Europa's benthos is similarly frigid, and life could probably evolve only in the warm chemistry possible around undersea volcanic cones and vents, if they exist.

Ganymede is another darling of science fiction. The largest moon in the Solar System, Ganymede is bigger than Mercury (5,262 km or 3,270 miles in diameter), but it has a gravity comparable to our more compact moon. Ganymede orbits Jupiter weekly at about 1,070,000 km (~665,000 miles), where it is safer, and radiation could be avoided in underground habitats on the side facing away from the planet. This would kill the view, but at least you would not die a victim of Jupiter's dynamo.

Ganymede's density, gravity, and tectonic features suggest three layers: a hard core, a slush rind, and a frozen skin. Ganymede has a weak magnetic field entwined with that of Jupiter's, probably generated by a metal core, as well as a trace O_2 atmosphere produced when water in the crust is split by cosmic rays. The hydrogen is lost to space, while the oxygen remains. Ganymede is cold (109 K), and its ocean, if liquid at all, is farther below the crust than even Europa's.

Callisto is NASA's choice for human exploration because it is just outside the Jovian radiation belts, and it has ample water (Troutman 2003). A shielded astronaut could walk on the surface. Callisto is a shade smaller than Mercury (4,820 km) and orbits Jupiter every sixteen days at roughly 1.88 million km (1.17 million miles). It is darker and absorbs more sunlight than Ganymede, making it slightly warmer. It has a thin CO_2 atmosphere, perhaps indicating dry ice on the surface. The crust may be 150 km (93 miles) thick but may still harbor a salty ocean.

It is not clear why NASA scientists like Callisto so much, other than as a place from which to launch people farther into the outer system. At 13 percent g, its escape velocity is similar to that of the Moon. Jupiter's gravity could also help hurl a ship home or onward toward Saturn.

The orbits of Jupiter and Saturn are 4.3 AU apart, roughly the distance from here to Jupiter. Our *Dawn*-like xenon ion drive ship would take nineteen years to go from Callisto to Saturn's Titan. From Earth, this is nearly eighteen Mars trips.

It would also be ill advised to approach Saturn too closely, for it too has a large magnetosphere. Saturn's magnetosphere is one-fifth the size of Jupiter's and similar in strength to our magnetosphere. It has two belts, the main and the so-called new, discovered by *Cassini* in 2004. These belts are interrupted by the magnificent rings, 170,000 miles across, pieces of ice ranging in size from grains to massive boulders. Like Jupiter, Saturn has many moons, but mighty Titan comprises 95 percent of the total mass of the moons.

Of Saturn's seven major moons, five are as big or bigger than Ceres, but only Iapetus and Titan are a safe distance from the planet and its magnetosphere. Titan is the second largest moon in the Solar System—also larger than Mercury. It orbits Saturn every sixteen days at 758,000 miles (~1.22 million km). Saturn would appear roughly ten times as large as the Sun does from here and the Sun roughly ten times as small.

Titan has an atmosphere denser than our own, with a surface pressure of 1.47 bars; it is 98 percent N_2 and smoggy. The smog is from methane and other hydrocarbons, and it acts as a radiation shield, but it also has an antigreenhouse effect, lowering Titan's temperature by reflecting sunlight away from the surface. Its temperature, colder than Saturn and confirmed in 2005 by *Cassini-Huygens*, hovers near 94 K ($-179°C$).

The density of Titan's atmosphere is high in part because of the cold, but it is still remarkable, considering that Mars has much stronger gravity. The atmosphere also contains argon, and the Ar^{40} isotope indicates that Titan's interior is warm, insulated perhaps by layers of methane and water ice that trap heat generated by Saturn's tidal forces. This could explain the volcanoes belching ammonia and methane, an important source of the atmosphere.

Titan was once thought to be covered by vast oceans of liquid methane, but astronomers now report broad sandy plains rich in organic tars called tholins and interspersed with liquid methane lakes. The plains are ridged in parallel and probably formed by atmospheric winds generated by Saturn's tidal forces (Lancaster 2006).

Huygens and *Cassini* provided evidence of two big hydrocarbon lakes on Titan in the high latitudes near the northern pole (Owen 2005). One measures some 100,000 square km (39,000 square miles), larger than any of the Great Lakes and comparable in size to the Caspian Sea. The other is larger than Lake Superior. Their properties suggest liquid "seas," and given Titan's atmosphere and temperature, they are likely mixtures of methane and ethane.

There is water on Titan too, but it is frozen at low temperature, so it is methane that cycles through Titan's atmosphere much like water does on Earth. The cycle exists because Titan's surface temperature of 94 K ($-179°$C) is between methane's melting (90 K; $-182.5°$C) and boiling (112 K; $-161°$C) points. This makes possible liquid methane rain. Precipitation has been observed in Titan's atmosphere by telescopes on Earth (Ádámkovics et al. 2007). As Titan rotates, the opacity of the troposphere measurably increases on the morning side of the leading hemisphere, where methane clouds condense to produce methane drizzle.

The moisture localizes to the equator over Titan's bright continent, Xanadu, and there, as dawn emerged, depending on the temperature and the humidity, you would be shrouded in cold methane rain for up to three Earth days. Like Earth, the rain would be influenced by topography and winds, akin to tropical systems on Earth. However, the clouds are thin, and models suggest misting, droplets of perhaps 0.1 mm, whereas a spring squall in Georgia produces raindrops ranging from 0.4 mm to 5 mm in diameter (Uijlenhoet et al. 2003).

An outpost on Titan would need nonsolar power far beyond twenty-first-century technology. The sand ridges suggest that wind power on Titan could be more effective than on Earth, because Titan's atmosphere is 50 percent denser than ours. However, Titan's winds measured during the *Huygens* descent were weak, in the range of one meter per second (~2.3 mph) (Bird et al. 2005).

Although Titan's protective atmosphere, ample water ice, natural gas, and other resources may seem tempting, Titan is not really a mini proto-Earth; it is so primitive that it is of no practical use unless it becomes the last resort of refugees from a roasting Sun.

MOONS OF THE ICE GIANTS

At Uranus and Neptune, things get even more difficult. Apart from Jupiter's dynamo, I disregarded cosmic radiation because at these distances and travel times, the ALARA principle breaks down; adequate shielding is the only option. Getting to Uranus, nineteen AU away, using our xenon ion drive would take another thirty-eight years. To avoid having to commit your great-grandchildren to space, let's switch from *Dawn* to a *New Horizons*–like ship that could reach Jupiter in a year. Saturn would then be two, Uranus four, and Neptune six years in deep space. Still, the shielding would have to be almost perfect.

The radiation is rivaled only by the cold, and as it gets colder, power requirements go up. The enormous electromagnetic dynamos of Jupiter and Saturn could hypothetically one day see a technology that could harness it, but capturing planetary radiation is only part of the solution; only giant robots could put the collectors in place and maintain them. Tapping into Jupiter's or Saturn's magnetosphere could be a huge boon on Callisto or Titan, respectively, but the technologies won't work at Uranus and Neptune. The twins have magnetic fields but not the formidable dipoles of Jupiter and Saturn (Stanley 2004).

Uranus and Neptune are each about four times as large as the Earth; Uranus is a bit larger, but it is less dense than Neptune. They are shrouded by hydrogen, helium, and a little methane, but like Jupiter and Saturn, they probably lack discrete surfaces. The moons are potentially more interesting, but they are colder than Titan, with no smog. Uranus's twins, Oberon and Titania, and Neptune's Triton are colder than liquid nitrogen.

Room temperature is 293 K (20°C), but Triton is 35 K. It is informative to see how much heat is needed to raise the temperature of ice on Triton by one Kelvin (1°C). The specific heat of ice is approximately 2,100 joules/kg K; thus, to bring a kilogram (2.2 lbs.) of ice from 35 K to its melting point, 0°C (273 K), we must add 500,000 joules of energy. Recall that *New Horizons* left

Earth with a radioisotope thermoelectric generator (RTG) that at Neptune will put out two hundred watts of power. At 25 percent efficiency, this would melt a kilogram of Triton's ice in four hours.

Some of this H_2O must be split by electrolysis to make O_2. Another back-of-the-envelope calculation shows that at 25 percent efficiency roughly eighty hours of RTG time would split a kilogram of H_2O. If 0.888 kg O_2 is recovered from this, it could support one astronaut for one day. In other words, a day's O_2 for an astronaut from ice on Triton would take more than three days of *New Horizon* power. In theory, O_2 could also be split catalytically from the ice without putting quite so much heat into it, but the calculation shows how hard it is to recover O_2 (and H_2O) from cryogenic ices.

Engineering texts are written about heat management in space, but the basics take only some arithmetic and a little familiarity with home insulation. The main difference between home and space is that the transfer of heat by moving fluids (air or water), called convection, is essentially absent in space (Griffin and French 2004). Most of the heat transfer in a near vacuum like space occurs by radiation and conduction.

A warm physical body will transfer heat to its colder surroundings until equilibrium is reached. Heat moves to the colder body in accordance with the second law of thermodynamics, and because the Sun is hotter than the Earth, we receive its heat by radiation. Heating at a distance occurs through the emission of electromagnetic waves depending on wavelength. The total radiation emitted from an object is found by the Stefan-Boltzmann law, a simple, useful relationship that requires us to know an object's surface area, its ability to radiate in the infrared region (called emissivity, ε), the tempera-ture difference between it and its surroundings, and Stefan's constant.

Because the Sun is hot, objects in our vicinity in its line of sight, such as satellites, easily overheat. A satellite passing into Earth's shadow is warmer than its surroundings, and the heat radiates into space. A bit of conduction occurs by direct molecular contact, for instance, from the satellite to mol-ecules in the surrounding medium, but this is slow because the molecules in space are far apart.

Aerospace engineers measure a satellite's surface area, analyze the radiant and conductive heat paths, and compute the solar load, internal heat genera-tion, and resistance to heat dissipation. In the end, however, they conduct a thermal vacuum test to simulate the hot and cold cycles the satellite will actually see in orbit. If the skin or the components overheat, the satellite is reconfigured. Once the prototype passes, it is ready for space.

The same principles apply on an ice moon. The warmest body, the Sun, transfers heat to it by radiation. The radiation raises the temperature by an amount that depends on distance from the Sun. The final temperature also depends on the object's surface area and emissivity (ε) compared to a perfect radiator, or *blackbody*. A true blackbody absorbs all the energy falling on it and radiates the maximum amount of energy possible at any temperature ($\varepsilon = 1.0$).

The equilibrium temperature of a habitat depends on the heat of the moon itself, its atmosphere, or both. If the atmosphere is windy, heat from the warmer object is carried away by convection. The habitat settles out at a temperature close to its surroundings, in thermal equilibrium with the Sun, the moon it sits on, the planet, and all other objects in its line of sight. A perfect blackbody on a line of sight with the Sun will not receive enough energy in the outer system to make it much warmer than the object it sits on, and heat must be added to raise its temperature into the tolerable range.

The addition of heat to a habitat to match the heat being lost to the environment puts it into *heat balance*. The heat lost by radiation and conduction are the most important. The radiation component is hard to compute because it is the sum of the heat gain from the Sun and other warmer bodies as well as losses to the surroundings.

Engineers can create thin, inexpensive, and efficient barriers to radiant heat loss, but the habitat will still slowly lose heat by conduction. Hence, other heat transfer characteristics of the habitat must be known, like the surface area and the material's thermal conductivity (k). Thermal conductivity is the rate at which heat is conducted away from a material of known thickness and area at a constant temperature difference over time. It is the power lost per unit area per degree, usually expressed in watts per meter per Kelvin (or °C).

We'll work with $1/k$, the resistance (R), because we want to prevent heat loss from a habitat, just like insulation reduces your home's heating bill in winter. The habitat's heat is determined by the area, the material's R value, and the temperature difference (ΔT) across the walls. This is just like the R value an insulation manufacturer puts on its products, except that international (SI) units are being used here, instead of British Thermal Units (BTUs).

Using this information, let's estimate the power needed to heat a habitat on Oberon. The temperature difference between Oberon (60 K) and our comfort zone is 233 K. The winter temperature at South Pole Station averages about 211 K ($-62°C$), about 150 K warmer than Oberon. To keep a one-inch-thick (2.54 cm) polyurethane-insulated sphere fifty feet in diameter at room temperature (293 K), the difference between the temperature inside and outside the sphere will be 233 K. Polyurethane is a great insulator and has low thermal conductivity (0.02 or, in our units, an R value of 50). Let us line the sphere with it ($4\pi r^2 = 7,854$ square feet). Dividing by R, multiplying by 233, and converting to kW, we need roughly 134 kW—more than the output of the eight-panel solar array of the ISS (84 kW).

Since the working area of the eight ISS solar panels is about three-quarters of an acre, about 1.2 acres would produce 134 kw. At twenty AU, the same array would occupy four hundred times the area for the same power. Thus, 480 acres of ISS solar array would heat a habitat fifty feet in diameter on Oberon. The thickness of the insulation can be increased, but this increases the mass and decreases the volume of the sphere.

Cryogenic ices or liquids also tend to degrade materials in contact with them. The entire field of cryogenic physics is devoted to understanding the effects of very low temperatures on matter. Long ago, my bespectacled high school chemistry teacher warned me in the usual way not to touch liquid N_2 or metals frozen by it by dropping a hot dog dipped in liquid N_2 onto the lab bench, where it shattered into pieces. Most plastics and metals also become brittle at very low temperatures. Plastics shrink at cryogenic temperatures and become more brittle as the temperature is repeatedly cycled, and cryogenic seals tend to fail. Structural integrity is affected by repeated cycles of pressure and radiation. This has been recognized for decades, for instance, in the degradation of the materials in nuclear reactors and even turned to advantage (in microchip etching), but it remains poorly understood (Wirth 2007). Today's best cold material is Teflon, which is stiff even at room temperature. The quest for flexible materials that seal and hold pressure at cryogenic temperatures is a highly underrated technological hurdle for both habitats and spacesuits.

There is a bright side to cryogenics; the possibility of using superconductivity—the loss of resistance to the flow of electricity—in space. The first compounds found to superconduct at temperatures above that of liquid nitrogen were discovered in 1986 by Bednorz and Muller in the IBM Zurich Research Laboratory. They won the 1987 Nobel Prize and caused a buzz because relatively little cooling was required. Since 1994, the record for superconductivity has stood at 138 K (at high pressures, this may reach 164 K). This means superconductivity can be put to use in the outer Solar System, for instance at Callisto.

"High-temperature" superconductors are ceramic sandwiches of oxides of copper arranged with one or more other elements. This complex arrangement makes the physics hard to study, which has delayed the development of materials for applications like limiting the power loss on long-distance, high-tension electrical lines. Some success has been had using YBCO wire (yttrium, barium, copper, and oxygen) to generate strong magnetic fields, but these tend to degrade the superconducting properties (Service 2006). A practical technology would be invaluable for habitat power, deflecting charged cosmic particles, mining operations, or containing fusion, if we decide to explore beyond the asteroid belt.

13. NEW STARS, NEW PLANETS

You may be scratching your head and wondering what's next. I have presented hard science and sprinkled it with bits of speculation, but now I must mainly speculate and sprinkle it with bits of science. You might want to visit another planetary system, but just in thinking about how to explore our own I have had to push our technology well into the future. Interstellar travel is notoriously intractable because of the limitations of physics (the speed of light) and of biology (the metabolic rate). This means no wormholes, quantum entanglement, tachyons, multiverses, or other exotic physics that does not yet reconcile with life and science. Whenever imagination bumps up against the limits of known science, brevity is best, unless you want to mislead and confuse someone.

We have discussed the shortcomings of chemical rockets, but there are also formidable problems with futuristic ion drives, nuclear fusion, and antimatter, too. Antimatter annihilates matter perfectly, releasing fantastic amounts of energy, and advanced ablative engines are being studied by NASA. An ablative engine vaporizes positrons (antielectrons) and electrons to produce thrust. The specific impulse of such an engine would be a dozen times that of the best chemical rocket, but the cost of producing positrons as a fuel for interstellar flight is impossibly high.

We need a faster ship, so I am simply going to invent one for our enjoyment. My ship, the *Cocoon*, goes 1 percent the speed of light (0.01 *c*) and could carry us to the edge of the Solar System or perhaps to another star. I could take you to Eris, twice as far away as Pluto, in a few months. Of course, given the distance, the radiation, and the cold, you might wonder why we are visiting this frozen hinterland. The only reason I can give you is that we would use it as a pit stop before disembarking for another star. This has nothing to do with saving time but with cosmic biogeography, which constrains us to hopping from island to island of resources.

Our ship must hold, harvest, and recycle O_2, H_2O, food, and other resources long enough to reach the next island. The sheer mass of the necessary resources is reason enough to depart from somewhere far from the Sun. Yet stars are light-years apart, a daunting reality trivialized by imaginary starships magically streaming through interstellar space at warp speed. Interstellar distances are, at the minimum, eight orders of magnitude (10^8) greater than the greatest voyage ever undertaken, the Earth to the Moon (2.56 light-seconds). Mars is under four light-minutes, and the Sun, 8.3 light-minutes. The next star, Alpha Centauri, is 4.36 light-years away—*276,000 times more distant than the Sun*. A fleet of space tugs laboriously nudging a Kuiper object, perhaps a cubewano rich in O_2, H_2O, CH_4, and other icy resources, from its orbital resonance with Neptune to harvest it during a millennial journey to another star might be a better way to go.

But for any multigenerational voyage, it will be a challenge to avoid a genetic bottleneck. When a small group of individuals is isolated reproductively from its original population, genetic diversity is reduced because some traits are no longer represented. The number of individuals needed to avoid this *founder effect* is unknown, but humans have passed through bottlenecks before, and our genetic diversity is lower than it is in the other great ape species. Population biologists have noticed genetic bottlenecks in many bird species in New Zealand. Using hatchability as a measure of fitness, it turns out that hatching failure is excessive among species passing through bottlenecks of fewer than 150 individuals but alleviated at a population of four times that size (Briskie and Mackintosh 2004). Assuming this would be the same for humans, a crew of six hundred is rather impressive for any starship. You would want a good reason to go and need a massive supply of O_2 and H_2O in cryogenic storage.

ON LEAVING THE SOLAR SYSTEM

I've asked my local experts to identify the edge of the Solar System for me, and they can't agree on precisely where it is, but any suitable ice ball at a thousand AU will do. A back-of-the-envelope calculation shows that our new ship could get there in 1.6 years, but even at 1 percent c, a brace of nuclear spaceships triple the size of Trident submarines traveling to Alpha Centauri would have to operate for 436 years on water ice electrolysis, and the reactor cores would have to be changed out ten times. Alpha Centauri is on the other side of a great interstellar ocean.

These distances are demoralizing to the prospective space traveler. Optimistic aerospace engineers don't foresee relativistic velocities anytime soon, even for unmanned spacecraft. However, let's ignore them and increase our velocity to 10 percent c (0.1 c, or 67 million mph, or 30,000 km/sec). This does not violate any laws of physics, and Alpha Centauri is suddenly only forty-four years away.

Alpha Centauri is a binary main sequence star (A and B), and a slightly closer red dwarf, Proxima or C, probably makes it a triplet. Star A, Rigel Kent, would seem familiar; it is a yellow G2 star of 1.1 solar masses. Alpha Centauri B is a smaller orange star. The system is older than the Sun, and Rigel Kent may run out of hydrogen in two hundred million years. But more to the point, Alpha Centauri has no planets.

The distance between the Sun and Alpha Centauri is typical of nearby stars (figure 13.1). In other words, the stars in our arm of the Milky Way are separated by about five light-years, which in the grand scheme of things is probably what made our location habitable. Eight of the twelve closest systems also contain M stars, or *red dwarfs*, of 0.1 to 0.6 solar mass and roughly half the Sun's temperature. They are also called *flare stars* because they show big spikes in intensity.

Red stars are far more common than yellow and orange stars, and they "live" a long time — one hundred to one thousand billion years, compared to the Sun's ten billion years. They account for three-fourths of the stars in the galaxy (brown dwarfs excluded). So M-star planets are worth a look, but the stars are faint, and it is hard to detect planets around them (Butler 2004).

The heat put out from most red stars would allow liquid water to exist only within the orbit of Mercury, and their planets are often tidally locked. The temperature of the hemisphere facing the star is far hotter than the one facing

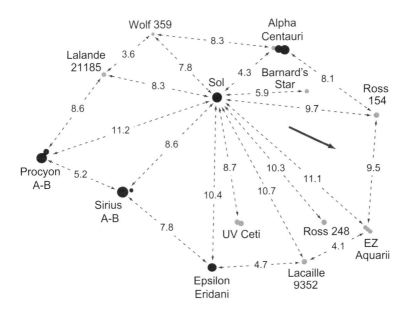

FIGURE 13.1. THE CLOSEST STAR SYSTEMS TO THE SUN AND THE DISTANCES AMONG THEM IN LIGHT-YEARS.
Most of the nearby stars are M-type stars or red dwarfs. The only known nearby planetary system is found at Epsilon Eridani. The heavy arrow (right) indicates the direction of the galactic center.

away from the star, even if heat is redistributed by an atmosphere or oceans. If red starlight is suitable for oxygenic photosynthesis and if the planet has a magnetic field, a thick atmosphere of the right gases, and the correct albedo, an M-star planet could be habitable. If not, it would be sterilized by solar flares unless life evolved underwater or underground.

Recall the red star Gliese 581 in chapter 2, whose system was anointed as the leading one for a future mission. In order to find out more about it, a probe must travel 20.4 ly (6.26 pc). At 0.1 c, it would not arrive for 204 years or return data for another 20.4 years. A 224-year return on investment is a tougher sell than Project Constellation, and a deep-space probe with a mean time-to-failure of more than two hundred years is itself an impressive feat.

Among the other nearby stars, only the orange Epsilon Eridani, 10.4 light-years away (3.2 pc), piques the interest of exobiologists, who envision it as a young Sun, the closest of its kind. It is famous, along with Tau Ceti, as being the first stars examined for signs of intelligent life by Frank Drake (b. 1930). Drake turned the eighty-five-foot radio telescope at Green Bank Observatory on it in 1960 but heard nothing. Decades later, a fledgling planetary system was discovered around it, and Drake repeated his famous experiment in 2010, but with the same negative result.

Epsilon Eridani, not yet a billion years old, is roughly 82 percent of the Sun's mass and one-third its luminosity. If an Earthlike planet did exist there, liquid water could be maintained within about 0.5 to 0.9 AU, the equivalent of Venus in our system. At that distance, a planet would not be tidally locked, and it would have a year similar to that of Venus. A "wobble" analysis of Epsilon Eridani in the 1990s suggested that it had a large planet (Benedict et al. 2006).

In 1998, a disk was photographed around Epsilon Eridani. It spread over a region thirty-five to seventy-five AU in size, which is similar to the forty to 120 AU range of our Kuiper belt. A clearing was seen around the star at about thirty AU, a distance equivalent to that of Neptune in our system, suggesting a planet was sweeping the dust from its orbit (Saumon et al. 1996).

The planet was thought to orbit the star at 7.7 AU and to be three times as large as Jupiter, but in 2000 astronomers pinned it to roughly 3.4 AU and 1.5 times the mass of Jupiter. The orbit is also not as wobbly as once thought, varying from 0.25 to 0.70 AU over about seven years. This planet one day may be seen by the Webb telescope, but because of its mass and eccentric orbit, the odds of there also being a small rocky planet inside Epsilon Eridani's HZ are low (Brogi et al. 2009). If the planet comes no closer to the star than 2.7 AU, there could be a small planet in the HZ.

Since we upgraded the *Cocoon* to fly at 0.1 c, it could reach Epsilon Eridani in 104 years (disregarding acceleration and deceleration). This is theoretically possible with modern nuclear technology but would require a series of reactors brought online at thirty- or forty-year intervals.

Exoplanets began to show up in 1995, when a Swiss team discovered the first one orbiting 51 Pegasus. By 2012, planet hunters had confirmed over seven hundred exoplanets, most by radial velocity or astronomy (Schneider et al. 2011). The early methods for detecting exoplanets were crude and found mostly Jupiter-sized or bigger giants orbiting their stars too closely for Earthlike planets to exist. The first system with an outer gas giant arrangement like ours was found in 2008, but five thousand light-years away (Gaudi et al. 2008).

In 2005, a so-called hot Jupiter (HD 189733 b) was found transiting a small yellow star sixty-five light-years away. It orbits its star every 2.2 days and is one of the first two planets to be observed by spectroscopy. It is tidally locked and extremely hot (>900 K) but shows steam in the atmosphere

(Tinetti et al. 2007). The star is too far away to reach, and the planet too large and hot to inhabit, but the detection of an exoplanet's atmosphere was a first.

The first photograph of an exoplanet was published in 2008: a huge planet around young Fomalhaut, twenty-five light-years away (Kalas et al. 2008). This planet is roughly three times Jupiter's mass and twelve times as far from Fomalhaut as Saturn is from Sol. Even though Fomalhaut is twice the Sun's mass and has a temperature of 8,400 K, this planet is surely an ice giant.

This portrait of gas giants, ice giants, or hot super-Earths in close orbits is of importance mainly to astronomers. The exoplanets mentioned so far are huge, and their gravitational fields are far greater than we can handle. The discovery of so many huge exoplanets represented a resolution bias in our detection methods against Earth-sized planets.

To address this issue, NASA's *Kepler* spacecraft was placed into an Earth-trailing orbit in 2009 to look for planets one-half to twice the size of the Earth in the habitable zones of their stars. Pointed at a single field in the Cygnus-Lyra region for 3.5 years, *Kepler* monitored one hundred thousand main-sequence stars for planets, using the transit method. The guess was that it would find roughly fifty Earth-sized planets that transit their stars about once a year.

Within a year, Kepler detected more than *two thousand* candidates, including many multiple-planet systems likely to include small rocky planets. Two such planets announced in December 2011 were orbiting Kepler-20, a G-type star; one was Earth-sized and the other about 13 percent smaller (Fressin et al. 2011). However, the planets are much closer to their star than Mercury is to the Sun, and Kepler-20 is 945 light-years away. At the 2012 American Astronomical Society meeting, three small exoplanets were announced orbiting a small red star, known as KOI-961, one-sixth the size of the Sun. The three exoplanets are 0.78, 0.73, and 0.57 times the size of Earth, and the smallest one is slightly larger than Mars. But they are 120 light-years away, and they, too, orbit their star very closely.

It is claimed that on average there is more than one planet per star in the Milky Way (Cassan et al. 2012), yet the goal of all this planet hunting is a nearby planet like our own, on which life may have evolved. Such a planet would not only be within the HZ of its star but also have reasonable gravity and an atmosphere containing O_2 and H_2O and no deadly gases. In the interest of Sagan's goal, we also would want to know beforehand if it was home to an inimical species best left alone. Many scientists have thought seriously about this over the years, even before the first spectral measurements were made on our own so-called pale blue dot from millions of miles away by the *Galileo* spacecraft (Sagan et al. 1993).

The spectrum of a planet or star is obtained by scanning a range of wavelengths for differences in the intensity of light emitted, reflected, or passing though the object or its atmosphere. The *Galileo* detected H_2O, O_2, and methane (CH_4), but subsequent studies of Earth light reflected back from the Moon during a lunar eclipse have provided a more accurate glimpse of how we would appear from deep space during a transit across the Sun—in other words, how we would look as an exoplanet.

Visible and near-infrared spectra of Earth obtained during one eclipse in 2008 indicated that light was being absorbed mainly by five molecules (Pallé et al. 2009): O_2, H_2O, CH_4, ozone (O_3), and CO_2. Notably, N_2 was absent, which has almost no spectrum, although its presence was inferred from its interactions with transient O_2-O_2 dimers. Thus, we know the essential spectral signature of our living planet. In contrast, when Venus transits the Sun, its atmospheric spectrum is very different. Venus shows intense absorption by CO_2 and a distinct lack of O_2 and H_2O. This is the signature of a hot, dry planet—no water and a runaway greenhouse effect.

Such information suggests that the presence of carbon-based life might be inferred from the transmission spectra of an exoplanet as it makes its way across its star. Specific atmospheric signatures could imply the absence of life, or the presence of anaerobic or aerobic life—or even technologically advanced life.

In the first two billion years of life on Earth, we had a reductive atmosphere of N_2, H_2O, CO_2, and CH_4, but no O_2. Life was anaerobic, and O_2 was toxic. Hydrogen (H_2) was absent because it escaped into space or was consumed by certain microorganisms for energy. Thus, an exoplanet rich in atmospheric H_2 may have nothing there consuming it, making for a low probability of life as we know it. On Earth, hydrogen is an important energy source for invertebrates that live in symbiosis with chemosynthetic bacteria near deep-sea hydrothermal vents (Petersen et al. 2011). The energy yield from the breakdown of H_2 for chemosynthetic growth is higher than for CH_4, sulfide (H_2S), or other simple electron donors. Thus, planets with water but no H_2 or O_2 might be the best candidates for anaerobic life.

After the Great Oxidation Event, O_2 appeared, and CO_2 levels rose in our atmosphere, while water and methane remained stable. Since O_2 oxidizes methane to CO_2, the presence of methane and O_2 advertises a renewable source of the gas—that is, *life*—contributing to our planet's distinctive signature. Of course, there are other atmospheric sources of methane, as on Titan, where the gas issues from cold volcanoes.

Finally, evidence of a synthetic gas in the atmosphere of an exoplanet, an unusual molecule unlikely to have been generated by natural processes, could suggest a technological civilization. Such a civilization might use synthetic chemistry to produce goods, releasing a telltale byproduct into the atmosphere, for instance chlorofluorocarbons, the notorious source of the expanding ozone hole on our own planet.

THE INTERSTELLAR MEDIUM

In thinking about traveling so far through interstellar space, perhaps it is possible to harvest the necessary resources. It has been said that if enough hydrogen (H) can be harvested from the interstellar medium (ISM), a fusion reactor could push the ship and generate O_2 and H_2O. This might provide both propulsion and resources for interstellar travel.

The ISM contains gas and dust in a ratio of about 99 percent gas and 1 percent dust (Croswell 1996). The gas is mostly hydrogen (90 percent) and helium (9 percent) in ionized and nonionized states, and the temperature ranges from very hot to very cold. Cold regions have a higher density; in some areas, clouds of hydrogen consisting of up to 10^{10} particles per cubic centimeter (cm^3) are found. Hot regions have densities as low as 10^{-4} particles per cm^3. On average, the ISM contains roughly one hydrogen atom per cm^3. This is nothing compared with air at sea level, which contains about 10^{19} molecules per cm^3.

The ISM is also lumpy and interspersed with colossal bubbles. We are currently passing through the hourglass-shaped Local Bubble, which is three hundred light-years across and has a hydrogen density of 0.07 atoms per cm^3, roughly fifteen times less than the Milky Way on average. Ignoring bubbles, a hypothetical fusion starship could use hydrogen fuel and make the other elements to support life. Fusion defines stars and recapitulates the chemical evolution of life.

The argument for a self-sustaining interstellar engine, the hydrogen ramjet, was laid out by the physicist Robert Bussard, who thought interstellar velocities close to the speed of light might be achieved using H from the ISM (Bussard 1960). Instead of carrying fuel, interstellar H would be scooped out by a ramjet to power the reactor. Bussard figured that 1 g of acceleration in a thousand-ton spacecraft in interstellar space could be achieved with a frontal scoop of ten thousand square kilometers (at 1 hydrogen atom/cm^3). Even using ultrathin materials, the scoop would vastly outweigh the ship. Furthermore, even in gas-rich regions the ship must reach 6 percent of c before the ramjet would work.

The concept was modified to ionize hydrogen and collect ions magnetically, but many physicists have still deemed it impractical because of the high power requirements of massive magnetic fields. And hydrogen collected as simple protons (H^+) requires more heat to fuse than do heavy isotopes like deuterium and tritium. Other types of ramjets release more energy than hydrogen, but they rely on rare elements. Catalytic engines that mix hydrogen and antihydrogen are more efficient and operate at lower temperatures than fusion, but generating and containing antimatter is well beyond today's technology.

Other elements in the ISM are even rarer, and this depends on local galactic chemical evolution, or GCE—chemistry on the grandest scale. These eternally slow processes deplete certain elements from the ISM by, for instance, depositing them onto dust grains. The next most abundant elements are O and C, found at just 180 and 150 atoms per million H. This "scoop factor" averages out to about one in six thousand.

Stars fuse H into helium (He) and He into other elements, releasing energy in the process of *nucleosynthesis*. Stellar furnaces also generate the elements of life, and stars a bit larger than the Sun make carbon from helium and beryllium, oxygen from carbon and helium, and the other elements up to atomic number 26, iron. Elements with higher atomic numbers are made by very slow neutron addition (the *s-process*), and are thus rare. Supernova explosions accelerate neutron addition (the *r-process*), producing elements of atomic numbers 27 to 92. Since fusion generates elements as heavy as iron, it makes most of the other elements of life too: C, N, O, P, S, and Ca. A few elements heavier than iron are necessary for life, like cobalt (^{27}Co), copper (^{29}Cu), zinc (^{30}Zn), selenium (^{34}Se), and iodine (^{53}I). These elements have abundances in space of roughly one ten-billionth of hydrogen.

Transmutation on a cottage scale is not trivial. Deuterium fusion produces helium "ash" that must be removed to avoid fouling the reactor, but hypothetically, this kills two birds with one stone: it powers and heats the ship, and the ash can be collected for other purposes, such as replacing O_2 lost from the ship. But all gases on a spacecraft trickle into space, including CO_2 and trace gases made by the body: carbon monoxide (CO), methane (CH_4), and nitric oxide (NO). Waste brine also contains ammonia, in the form of urea—$(NH_2)_2CO$—and materials lost by the oxidation of metals and plastics must also be replaced. In other words, everything is on the table.

To use fusion to generate O_2, helium is converted to carbon, which requires a very high temperature. Suppose our reactor can convert every bit of the harvested hydrogen to helium by harnessing the proton-proton ("p-p") reaction, just like the Sun. Very hot stars actually "burn" helium (^{4}He) into carbon (^{12}C), by the *triple alpha* process. Triple alpha refers to the fusion of three alpha particles (helium nuclei):

$$^4He + {}^4He \rightarrow {}^8Be \ \& \ {}^8Be + {}^4He \rightarrow {}^{12}C + \text{gamma radiation}$$

When the temperature reaches about 2×10^8 K, carbon plus helium produces oxygen:

$$^{12}C + {}^4He \rightarrow {}^{16}O$$

The conversion rates in massive stars are indeterminate, but C- and O-rich stars have different fates, ending as neutron stars or black holes, respectively (Gai 2003). If we generate equal amounts of C and O (C/O = 1), recover them fully, and harvest all the H and He in one cubic kilometer of average ISM on a ship traveling at 0.9c, enough O_2 could be made to support one person for one day, if we are recycling it at 95 percent efficiency.

Stringing processes together also rapidly degrades conversion efficiency. Imagine two three-step processes designed to make glucose from H_2O and CO_2. If each three-step process has an efficiency of 95 percent, the overall efficiency for glucose production is 0.95 x 0.95, or about 90 percent. If each of the six steps is 95 percent efficient, the overall efficiency would be $0.95^3 \times 0.95^3 \times 100 = 73.5\%$. Molecules that do not end up as glucose must be recovered, too. Thus ISM harvesting depends on extremely high velocities, a scoop the size of the island of Puerto Rico, a perfect fusion reactor, and unsustainable recycling efficiencies.

CHANGING OURSELVES

Considering how to get around such seemingly insurmountable limits often leads science fiction authors into fantasizing about changing ourselves. But genetic engineering is subject to ethical issues and is still a nascent science, so discussing it here would veer too far into pure speculation. Similarly, no matter how fascinating a topic it is, let's also set artificial intelligence aside for now.

What about conserving energy on a long space voyage by sleeping? Let's loosen the definition of "sleep" to include hibernation, estivation, hypothermia, and suspended animation because normal sleep won't conserve enough energy. The hypothetical limit would be lowering the body's energy requirements to zero by placing it in equilibrium with its environment. The crew would effectively be dead until the ship arrives at a new destination, where they would be resurrected. This has been proposed, for example, by using an atomic clock to time the delivery of a "Lazarus cocktail" when the ship arrives at a habitable exoplanet a million years later. By "freezing" biology out, this suggests that travel to anywhere in the universe is possible.

Close scrutiny, however, quickly uncovers two fatal flaws in this scheme. First, reanimation involves two regulated processes—a clock and a cocktail. The clock, the cocktail, and the people are present throughout the voyage and subject to random events. Even for a perfect system, a rare catastrophe—a comet, a burst of radiation, a neutron star, a supernova, or black hole— would destroy the clock, the cocktail, and the people. The galaxy is rife with cataclysmic events, and the probability of disaster befalling a ship on a protracted interstellar voyage approaches 1.

Suppose ten thousand light-years away, in an adjacent arm of the Milky Way, we find a beautiful water planet with an O_2-rich atmosphere orbiting an orange star. The *Cocoon* sets out on a hundred-thousand-year trip to this planet. We predict that a million-year event would destroy our clock one time out of ten and that the ship should get through nine times out of ten. This is far better than the success rate of any Russian Mars mission. The trouble is that in a galaxy of two hundred billion stars, galactic cataclysms occur far more often than once in a million years.

Massive stars, for instance, explode after about a million years. These supernovas can be tracked by the amount of radioactive ^{26}Al released as they die. The galactic ^{26}Al content has been used to estimate the supernova occurrence rate in the Milky Way, and the predicted event rate is roughly 1.9 per century (Diehl et al. 2006). In one hundred thousand years, the *Cocoon* would "see" 1,900 supernova explosions, and chances are that one would be close enough to destroy it.

Changing ourselves does not violate the laws of physics. Carbon and O_2 utilization rates may be reduced, perhaps even by a hundred-fold, but even under anesthesia and hypothermic arrest, the metabolic rate does not reach zero. Metabolic debt accumulates, and prolonged periods of deep hypothermia cause deficits in brain function (Hogue et al. 2006). Even if metabolic suppression worked, suspending animation, as we just saw, lacks tractability.

Just to be sure, consider *hibernation* or estivation (another form of torpor), where body metabolism falls to conserve energy. Hibernation involves profound changes in physiology and behavior associated with sleeping and burning fat (Carey et al. 2003). Although some hibernating mammals can lower their body temperature to 0°C and their metabolic rates to 1 percent of normal, this saves less energy than you think. The survival advantage is conferred by the animal not having to forage for food when it is scarce. The hibernator makes hay in the fall, fattening up in order to survive a cold winter.

Hibernating mammals also awaken at various intervals, and body temperature rises into the normal range. According to Dr. Brian Barnes, director of the Institute of Arctic Biology in Fairbanks, these episodic arousals in large hibernators may paradoxically be needed to prevent sleep deprivation. Thus, in the black bear, hibernation is not actual sleep, which seems to occur during its arousals. Black bears also inexplicably maintain normal muscle and bone mass over the winter despite losing up to one-third of their body weight. Understanding hibernation in the black bear would be of great interest to NASA. This situation is not favorable for space travel (Piantadosi 2003). And unlike calorie restriction, hibernation does not seem to prolong lifespan (Gredilla and Barja 2005, Colman et al. 2009).

Could genetic engineers one day modify people for space travel? This question mirrors the revolution in genetic medicine touted to enhance everything from intelligence to longevity. Why not design miniature astronauts with low metabolic rates? But this is no match for the limits demonstrated by the clock and the cocktail.

For nonhibernators, the limit for reducing metabolic rate is a factor of about one-half before cardiac and brain cells begin to die. Reducing body temperature by 10°C (to 27°C) halves the metabolic rate, but people are then near coma and cannot rescue themselves from the cold (Danzl and Pozos 1994). Apart from mild hypothermia, which is induced to protect temporarily an ischemic brain, body cooling has not proven useful in medicine. At a metabolic rate of 1 percent of normal, there is insufficient energy for physiological homeostasis.

In zebra fish and mice, body temperature and metabolic rate can be cut using small amounts of hydrogen sulfide (H_2S), better known as sewer gas (Blackstone et al. 2005). This idea received press for the notion that small amounts of H_2S might induce metabolic torpor without tissue damage. However, metabolic inhibitors eventually wreck cellular machinery.

It has been argued that reducing metabolism and body mass together would optimally restrict energy use. A tenfold reduction in an adult's body size, on a growth chart, yields the size of an average two-year old. At this age, the body is normal, but the brain and the reproductive system are not fully developed. Setting the ethical objections aside, molecular biology may coerce those activities to maturity without growth, giving us child-sized astronauts. However, there is a problem here, too.

The metabolic rate is set by a scale factor involving cell size, cell number, the nutrient network (microcirculation), support structures, and function. It varies most for tiny animals, like insects, and the intraspecies range is high (Chown et al. 2007). The range of metabolic rates is also high. Our metabolic rate is near the top, and our body mass scales to the 0.75 power ($M^{0.75}$). This is also reflected in the enzymes of respiration and the mitochondrial density of cells. Tiny mammals such as the bumblebee bat and the Etruscan shrew have the highest metabolic rates per gram and more efficient scale factors than we do. They operate near the maximum thermodynamic efficiency of respiration (Dobson and Headrick 1995). In other words, a smaller body requires a higher metabolic efficiency, and the limits have already been tested by nature.

SCIENCE FACT, SCIENCE FICTION

The science of limits has indicated that the idea of huge multigenerational starships dragging cubewanos along with them, although conceivable, pushes the envelope for even nearby stars not because of technology but because the ratios of distance over velocity and energy over mass are so high. We may get a little boost by harvesting energy from the ISM, but our galaxy is not simply, as astronomers assert, our own backyard. It is 100,000 light-years across with a central bulge 16,000 light-years in diameter. The Milky Way packs in perhaps 200 billion stars, and our spot in the Orion Arm is roughly 27,000 light-years from the center.

The enormity of the galactic disc always brings out the old science fiction trick of getting extra mileage out of relativity. Relativity sets the speed of light as a hard restriction on velocity, a disadvantage. But clocks run slower and time "dilates" with velocity, which is an advantage. To see this, let's accelerate to 0.9 c with a giant ding-a-ling reactor that operates flawlessly for millennia.

The speed limit c is offset by time dilation. The easiest way to understand time dilation is to realize that time is shortest at "rest." This rest time is called "proper time." Since the math has been done many times, we know that time dilates about 2.3-fold relative to proper time on a ship traveling at 0.9 c. A factor of 2.3 means our ship's resources are consumed at only 44 percent the original rate, giving it a greater range.

By increasing the velocity of the *Cocoon* to 0.9 c, which had a range of ten light-years at 0.1 c, it is nine times faster, and we use roughly 2.3 times less resources. This means the resources we originally needed for Alpha Centauri (4.4 ly) would now take us to Epsilon Eridani (10.4 ly). Instead of 104 years, Epsilon would take only 11.7 years. If we ran the ship provisioned for 104 years up to 0.9 c, our range would improve to 215 ly, but that still keeps us within our arm of the galaxy.

At 0.99 c, the time factor increases to seven, but every step proportionally increases the mass of the ship and the energy required to accelerate it. In nature, only ultra-high-energy galactic rays are imbued with such high velocities, typically *single protons* generated by massive black holes at violent galactic centers. These active galactic nuclei (AGN) spew particles across the universe (Pierre Auger Collaboration et al. 2007).

Tinker with this more on your own, but don't be surprised, as some of my students have been, that near-perfect harvesting and recycling on a ship, large or small, traveling at 0.99 c won't get tiny astronauts with drastically low metabolic rates to Andromeda alive. It illustrates why biologists find interstellar travel fantastic and why physicists concoct suppositious physics. The relativity game simply trades mass and energy for time. Even extremely rare hypervelocity stars, those capable of escaping the Milky Way, travel at less than 0.005 c.

Although I set a discussion of cybernetic organisms aside earlier, the problems we have discussed in this chapter led John Von Neumann and others to propose constructing "indestructible" life forms with artificial intelligence and miniscule energy requirements to act as interstellar probes. These probes would actively self-navigate and self-correct, perhaps using cosmic radiation for energy. They could also hibernate when passing through bubbles of nearly empty space. Energy would be stored when available and consumed when not. Like biological organisms, however, robots must be impervious to radiation, to deep thermal cycles, and to the other hazards of space.

There is also no opposition to programming "beings" into trillions of "hardened" spores and disseminating them into space to germinate when conditions are right. Such directed panspermia violates none of our rules, and hypothetically, the right blueprint could allow life to evolve elsewhere and equip it with information. If we sent out trillions upon trillions of blueprints, one might find a home. It might seem that this might be a good plan to implement in and for the far future, but sending spores off to other stars and galaxies is strangely quixotic: they vanish, nothing is ever communicated to those who originally sent them, and life made by blueprint may never learn of—or believe in—its (artificial) origin. This is the cosmic teapot in reverse, fantasized about by those trying to escape understanding the evolution of life on our planet.

Assuming that anything imaginable can be done, science fiction writers also envision civilizations that move star systems. Indeed, the Sun moves toward Vega in a giant ellipse around the galactic center (Garlick 2002), making a complete circuit—a galactic year—once every 250 million years. This is roughly one parsec every 4,500 years, and we have moved about a parsec relative to the galactic center since the beginning of recorded history. The planet was last here during the time of Pangaea and the Permian-Triassic extinction, the so-called Great Dying.

Alas, Von Neumann was right all along, unless Einstein overlooked something or the quantum mechanics guild learns to fold space. In the meantime, new horizons involving probes and people must be set by practicable and cost-effective science. We must be patient as our horizons change because time, distance, cost, risk, and the list of choices increase, making decisions about what's next more difficult.

Perhaps it is not quite fair to have set up thought experiments using spacecraft technology that always lags just behind what is needed to reach the next island. This is the paradox: the ratio of suitable to total alternatives shrinks as our vision expands. Not long ago, we inhabited one planet of eight. Now we're on one of thousands, and next we'll be on just one of millions. Most are wildly exotic, and so far, none resembles ours. Our own uniqueness and space's insuperability are the best incentives we have to take the best possible care of Spaceship Earth.

NOTES

1. MEN AND MACHINES

1. By early 2014, Space X's had met its requirements as a critical cargo and scientific equipment to the ISS under a $1.6 billion Cargo Resupply Services contract with NASA. SpaceX also unveiled the Dragon 2/Falcon 9 system, which can carry a crew of seven into orbit and is designed to be refueled and reused. Its ability to land under power would solve a key problem for round-trip missions to the Moon and Mars and offers the first real opportunity for the commercialization of space at a reasonable cost. A dependable Dragon 2/Falcon 9 could pave the way for private enterprise to lead us back to the Moon and to Mars in the next twenty years.

5. BACK TO THE MOON

1. In 2014, Russia too added a manned lunar mission to its docket for after 2025. To paraphrase the head of strategic planning for Roscosmos: "The Moon is an interesting object if one takes the next step and builds a manned station there. It has

everything, including water to provide the station with products vital for life—rocket fuel, building materials, everything." However, the Russians are unlikely go it alone, especially if the bite of economic sanctions imposed by the West over their Ukraine policies continues for very long. Hence, the other space stakeholders are still drawn to lunar exploration, while NASA quixotically seems intent on saving the world from an unidentified killer asteroid.

12. BIG PLANETS, DWARF PLANETS, AND SMALL BODIES

1. NASA scientists have unveiled a plan to move a small asteroid (about eight meters wide) into close orbit around the Moon. This so-called Asteroid Redirect Mission (ARM), dropped out of the Asteroid Initiative, has two goals: it would test new propulsion technology to nudge potentially hazardous asteroids out of the Earth's path, and it would test "new technologies" needed for a manned Mars mission.

   If we examine ARM a little more closely, we find a list of candidates drawn up by scientists at the Goddard Space Flight Center on the Near-Earth Object Human Space Flight Accessible Targets Study (NHATS). NHATS located some 1,200 potential targets for the ARM mission and computed trajectories for hypothetical missions to each from Earth. In their existing locations, the energy cost and round-trip spaceflight time to each would be less than for a Mars mission. But NASA hopes to capture one in a "bag" and drag it into a metastable orbit around the Moon. Then a team of astronauts, carried by the new SLS and Orion spacecraft, would make a visit sometime after 2020. Apart from the fact that bagging a tiny Earth-trailing asteroid is an entirely different matter than changing the orbit of a kilometer-sized, Earth-crossing asteroid traveling at 50,000 mph, it is important to ask whether this is the best use of our limited manned spaceflight resources. Although ARM would again take people beyond the Van Allen belts, allowing us to learn more about the deep space radiation, which is critical for Mars, the astronauts would still spend their time in microgravity, and won't need (and hence, we won't be developing) the surface technologies for Mars planetary exploration—our real objective. Even if ARM succeeds, these surface technologies must still be developed, which could require an additional twenty years. Until then, any mission to Mars would be relegated to a stop at Deimos or to a one-way trip followed by lengthy, expensive, and risky resupply missions.

BIBLIOGRAPHY AND ADDITIONAL

READING

INTRODUCTION AND CHAPTER 1

Aldridge, Edward C., et al. 2004. "Report of the President's Commission on Implementation of United States Space Exploration Policy: A Journey to Inspire, Innovate, and Discover." Washington, D.C.: U.S. Government Printing Office.

APS (APS Panel on Public Affairs). 2004. "The Moon-Mars Program." American Physical Society. November. http://www.aps.org/public_affairs/index.cfm.

Brumfiel, G. 2007. "Where Twenty-four Men Have Gone Before." *Nature* 445: 474–478.

CBO (Congressional Budget Office). 2004. "A Budgetary Analysis of NASA's New Vision for Space Exploration." Congress of the United States, Superintendent of Documents. Washington, DC: U.S. Government Printing Office. September.

CAIB (Columbia Accident Investigation Board). 2003. *Final Report on Columbia Space Shuttle Accident.* Vol. 1. Washington, D.C.: U.S. Government Printing Office.

Cucinotta, F., et al. 2002. "Space Radiation Cancer Risk Projections for Exploration Missions: Uncertainty Reduction and Mitigation." NASA TP 2002-210777. Springfield, Va.: National Technical Information Service.

Fuller, Steve. 2004. *Kuhn vs. Popper: The Struggle for the Soul of Science.* New York: Columbia University Press.

Klotz, I. 2006. "NASA Gambles All for a Shot at the Moon." *New Scientist* 189 (January 14): 8–9.

Krupp, N. 2007. "New Surprises in the Largest Magnetosphere of our Solar System." *Science* 318:216–217.

Lawler, A. 2002. "Technology Is Essential but It's a Tough Sell." *Science* 295:39.

Lawler, A. 2007. "Lunar Science: Asking for the Moon." *Science* 315:1482–1484.

Lawler, A. 2009. "Trouble on the Final Frontier." *Science* 324:34–35.

McComas, D. J. et al. 2007. "Diverse Plasma Populations and Structures in Jupiter's Magnetotail." *Science* 218:217–220.

NASA. 2004a. "The Vision for Space Exploration." 20546NP-2004-01-334-HQ. Washington, D.C.: NASA Headquarters. (electronic version at www.nasa.gov). February.

NASA. 2004b. *NASA's Implementation Plan for Space Shuttle Return to Flight and Beyond.* Vol. 1, rev. 2.2. Houston, TX: Johnson Space Center, (electronic version at www.nasa.gov). August 27.

NASA. 2007. "How We'll Get Back to the Moon." http://www.nasa.gov/mission_pages/constellation/main/cev.html.

NASA. 2011. "Human Space Exploration Framework Summary." (HEFT) Final Brief.

New Atlantis. 2005. "Relaunching NASA: Back to the Moon by 2018—or Sooner." Editorial. *New Atlantis* 10 (Fall): 118–122.

Piantadosi, C. A. 2006. "Lunar Exploration and the Advancement of Biomedical Research: A Physiologist's View." *Aviation and Space Environmental Medicine* 77: 1084–1086.

Scafetta, N. 2012. "Testing an Astronomically Based Decadal-Scale Empirical Harmonic Climate Model vs. the IPCC (2007) General Circulation Climate Models." *Journal of Atmospheric and Solar-Terrestrial Physics.* doi:10.1016/j.jastp.2011.12.005.

Schilling, G. 2006. "Long Trek to Solar System's Last Frontier Begins." *Science* 311:172.

Shiga, D. 2009. "At the Crossroads, Tryin' to Flag a Ride." *New Scientist* 202 (2705): 6–7.

Shiga, D. 2011. "The Space Race Takes Off Again." *New Scientist* 211 (2820): 6–7.

SCTG (Stafford-Covey Return to Flight Task Group). *Final Report of the Return to Flight Task Group Assessing the Implementation of the Columbia Accident Investigation Board Return-to-Flight Recommendations.* Houston, Tex.: NASA.

CHAPTER 2

Beall, C. M. 2007. "Two Routes to Functional Adaptation: Tibetan and Andean High-Altitude Natives." *Proceedings of the National Academy of Sciences of the United States* 104:8655–8660.

Boulding, K. E. 1966. "The Economics of the Coming of Spaceship Earth." In *Environmental Quality in a Growing Economy*, ed. H. Jarrett. Baltimore, Md.: Johns Hopkins University Press.

Brockmann, D., L. Hufnagel, and T. Geisel. 2006. "The Scaling Laws of Human Travel." *Nature* 439:462–465.

Cannon, W. B. 1932. *The Wisdom of the Body*. New York: Norton.

Fregly, M. J., and C. M. Blatteis, eds. 1996. *Environmental Physiology*. 2 vols. New York: Oxford University Press.

Johnson, D. L., B. C. Roberts, and W. W. Vaughan. 2002. *Reference and Standard Atmosphere Models*. Huntsville, Ala.: NASA; Springfield, Va.: Marshall Space Flight Center, National Technical Information Service.

Macrae, Norman. 1999. *John von Neumann: The Scientific Genius Who Pioneered the Modern Computer, Game Theory, Nuclear Deterrence, and Much More*. Providence, R.I.: American Mathematical Society.

Margot, J. L., et al. 2007. "Large Longitude Libration of Mercury Reveals a Molten Core." *Science* 316:710–714.

NASA. 2003. Procedures and Guidelines NPG: 8705.2 Human-Rating Requirements and Guidelines for Space Flight Systems Effective Date: June 19, 2003 Expiration Date: June 19, 2008. Q/Office of Safety and Mission Assurance.

Stott, P. A., D. A. Stone, and M. R. Allen. 2004. "Human Contribution to the European Heat Wave of 2003." *Nature* 432:610–614.

Udry, S., et al. 2007. "The HARPS Search for Southern Extrasolar Planets, XI. Super-Earths (5 and 8 $M_\oplus$) in a 3-Planet System." *Astronomy and Astrophysics* 469: L43–L47.

Yi X., et al. 2010. "Sequencing of 50 Human Exomes Reveals Adaptation to High Altitude." *Science* 329:75–78.

CHAPTER 3

Amundsen, Roald. 2001. *The South Pole: An Account of the Norwegian Antarctic Expedition in the Fram, 1910–12*. New York: Cooper Square Press.

Baader, Gerhard, et al. 2005. "Pathways to Human Experimentation, 1933–1945: Germany, Japan, and the United States." *Osiris*, 2nd ser. 20:205–231. http://links .jstor.org/sici?sici=0369.

Campbell, M. R., et al. 2007. "Hubertus Strughold: The Father of Space Medicine." *Aviation, Space, and Environmental Medicine* 78: 16–719.

Dill, D. B. 1979. "Case Histories of a Physiologist: F. G. Hall." *The Physiologist* 22:8–21.

Droessler, E. G., J. M. Lewis, and T. F. Malone. 2000. "Lloyd Berkner: Catalyst for Meteorology's Fabulous Fifties." *Bulletin of the American Meteorological Society* 81:2963–2973.

Gagge, A. P. 1986. "The War Years at the Aeromedical Lab: Wright Field (1941–46)." *Aviation, Space, and Environmental Medicine* 57:A6–12.

Guttridge, Leonard F. 2000. *Ghosts of Cape Sabine: The Harrowing True Story of the Greely Expedition*. New York: Berkeley.

Luedecke, C. 2004. "The First International Polar Year (1882–83): A Big Science Experiment with Small Science Equipment." *Proceedings of the International Commission on History of Meteorology* 1 (1): 55–64.

Malashenkov, D. C. 2002. IAF abstracts, Thirty-fourth COSPAR Scientific Assembly, the Second World Space Congress, October 10–19, Houston, Tex. P. IAA-2-2-05IAF.

Needell, A. 1992. "From Military Research to Big Science: Lloyd Berkner and Science-Statesmanship in the Postwar Era." In *Big Science: The Growth of Large-Scale Research*, ed. P. Galison and B. Herly, 290–311. Stanford, Calif.: Stanford University Press.

Nicolet, M. 1982. "The International Geophysical Year 1957/58." *Bulletin of the World Meteorological Organization* 31:222–231.

Piccard, A. 1997. "My Beautiful Air-Tight Cabin." In *From the Field: A Collection of Writings from* National Geographic, ed. C. McCarry, 106–108. Washington, D.C.: National Geographic Society.

Salmon, R. 2007. "The Scope of Science for the International Polar Year 2007–2008." http://www.ipy.org.

Schiermeier, Q. International Polar Year: In from the Cold. *Nature* 457:1072–177, 2009

Schley, W. S. 1885. *Rescue of Greely*. New York: Charles Scribner's Sons.

USAF School of Aerospace Medicine. 1950. *German Aviation Medicine, World War II*. Washington: Dept. of the Air Force.

Vorenburg, S. 2006. "Museum Removes Nazi Honoree: Space Researcher Dropped from Hall of Fame." *Albuquerque Tribune* (May 17).

Wieland, P. O. 1994. "Designing for Human Presence in Space: An Introduction to Environmental Control and Life Support Systems." NASA RP-1324. Huntsville, Ala.: Marshall Space Flight Center.

Weindling, Paul J. 2004. *Nazi Medicine and the Nuremberg Trials: From Medical War Crimes to Informed Consent*. Basingstoke: Palgrave Macmillan.

Wolfe, Tom. 2001. *The Right Stuff*. New York: Bantam.

CHAPTER 4

Columbia Accident Investigation Board (CAIB). 2003. Final Report on Columbia Space Shuttle Accident. Vol. 1. Washington, D.C.: Superintendent of Documents, U.S. Government Printing Office.

Compton, W. David. 1989. *Where No Man Has Gone Before: A History of Apollo Lunar Exploration Missions*. Special Publication-4214 (NASA History Series). Washington, D.C.: Scientific and Technical Information Office, NASA.

Dick, Steven J., et al., eds. 2007. *America in Space: NASA's First Fifty Years*. New York: Abrams.

Feynman, Richard P. 1988. *"What Do You Care What Other People Think?" Further Adventures of a Curious Character*. New York: Norton.

Hale, Edward E. 1869. "The Brick Moon. From the Papers of Colonel Frederic Ingham." *Atlantic Monthly* 24 (144): 451–460; (145): 603–611; (146): 679–689.

Lovell, James A. 1975. "Houston, We've Had a Problem." In *Apollo Expeditions to the Moon*, ed. E. M. Cortright. Special Publication 350. Washington, D.C.: Scientific and Technical Information Office, NASA.

NASA. 1970. Report of Apollo 13 Review Board. National Aeronautics and Space Administration Apollo 13 Review Board. June 15.

NASA. 1997. A History of U.S. Space Stations. IS-1997-06-ISS-009JSC. Houston, Tex.: Lyndon B. Johnson Space Center.

NASA. 2004. NASA's Implementation Plan for Space Shuttle Return to Flight and Beyond. Vol. 1, rev. 2.2. Houston, Tex.: Johnson Space Center.

NASA. 2005. Final Report of the [Stafford-Covey] Return to Flight Task Group Assessing the Implementation of the *Columbia* Accident Investigation Board Return-to-Flight Recommendations. Washington, D.C.: NASA.

National Research Council (NRC). 2004. Stepping Stones to the Future of Space Exploration. A Workshop Report. Aeronautics and Space Engineering Board. Washington, D.C.: National Academies Press.

National Research Council Committee on the Origins and Evolution of Life (NRCCOEL). 2003. Life in the Universe: An Assessment of U.S. and International Programs in Astrobiology. Washington, D.C.: National Research Council.

Rogers Commission. 1986–1987. Report of the Presidential Commission on the Space Shuttle *Challenger* Accident (Rogers Commission Report), June 1986; and Implementations of the Recommendations, June 1987. Washington, D.C.: U.S. Government Printing Office.

Siddiqi, A. A. 2000. *Challenge to Apollo.* Washington, D.C.: NASA.

Sobel, Dava. 1996. *Longitude: The True Story of a Lone Genius Who Solved the Greatest Problem of His Time.* New York: Walker.

Wieland, P. O. 1998. *Living Together in Space: The Design and Operation of Life Support Systems on the International Space Station.* Vol. 1. NASA TM/ 98 206956. Springfield, Va.: National Technical Information Service.

CHAPTER 5

Borra, E. F., et al. 2007. "Deposition of Metal Films on an Ionic Liquid as a Basis for a Lunar Telescope." *Nature* 447:979–981.

Clery, D. 2007. "For Extreme Astronomy, Head Due South." *Science* 315:1523–1524.

Cohen, David. 2007. "Earth Audit." *New Scientist* 194 (2605): 34–41.

Colaprete, A, et al. 2010. "Detection of Water in the LCROSS Ejecta Plume." *Science* 330:463–468.

Crow, J. M. 2011. "Unsung Elements." *New Scientist* 210 (2817):37–42.

Dalcanton, J. J. 2009. "Eighteen Years of Science with the Hubble Space Telescope." *Nature* 457:41–50.

Duke, M. B., et al. 2003. *Lunar Surface Reference Missions: A Description of Human and Robotic Surface Activities.* Houston, Tex.: NASA Johnson Space Center Houston. Available from the NASA Center for Aerospace Information (CASI) 7121 Standard.

Georg, R. B., et al. "Silicon in the Earth's Core." *Nature* 447:1102–1106.

Gladstone, G. R., et al. 2010. "LRO-LAMP Observations of the LCROSS Impact Plume." *Science* 330:472–476.

Gott, R. J. 1993. "Implications of the Copernican Principle for Our Future Prospects." *Nature* 363:315–319.

Gott, R. J. 2007. "Why We Must Leave Earth." *New Scientist* 194: 51–54.

Hanover, M. D., et al. 2006. "The James Webb Space Telescope." *Space Science Review* 123:485–606.

Hartmann, W. K. 1997. "A Brief History of the Moon." *The Planetary Report* 17:4–11.

Haruyama, J., et al. 2008. "Lack of Exposed Ice Inside Lunar South Pole Shackleton Crater." *Science* 322:938–939.

Knaapen, A. M., et al. 2004. "Inhaled Particles and Lung Cancer. Part A: Mechanisms." *International Journal of Cancer* 109:799–809.

Liu, Y., et al. 2008. "Characterization of Lunar Dust for Toxicological Studies. II: Texture and Shape Characteristics." *Journal of Aerospace Engineering* 21:272–279.

Mann, A. 2011. "Scope Sails Into Budget Void." *Nature* 468:353–354.

McKay, D., et al. 1991. "The Lunar Regolith." In *Lunar Sourcebook*, ed. G. Heiken et al., 285–356. Cambridge: Cambridge University Press

Murdin, Paul, ed. 2001. *Encyclopedia of Astronomy and Astrophysics*. Bristol: Institute of Physics Publishing.

National Research Council Committee on the Scientific Context for Exploration of the Moon (NRCCSCEM). 2006. *The Scientific Context for Exploration of the Moon: Interim Report.* Washington, D.C.: National Academies Press.

Park, J., et al. 2008. "Characterization of Lunar Dust for Toxicological Studies. I: Particle Size Distribution." *Journal of Aerospace Engineering* 21: 265–271.

Pieters, C. M., et al. 2009. "Character and Spatial Distribution of OH/H2O on the Surface of the Moon Seen by M3 on Chandrayaan-1." *Science* 326:568–572.

Sanderson, Katharine. 2007. "The Sunniest Spot on the Moon. SMART-1 Data Indicates a Good Spot for a Lunar Base." *Nature News*. doi:10.1038/news.2007.182.

Schmidt, Harrison H. 2006. *Return to the Moon: Exploration, Enterprise, and Energy in the Human Settlement of Space.* New York: Springer.

Sharpe, B. L., and D. G. Schrunk. 2003. "Malapert Mountain: Gateway to the Moon." *Advances in Space Research* 31:2467–2472.

Sunshine, J. M., et al. 2009. "Temporal and Spatial Variability of Lunar Hydration as Observed by the *Deep Impact* Spacecraft." *Science* 326:565–568.

West, J. B. 1998. *High Life: A History of High Altitude Physiology and Medicine.* New York: Oxford University Press.

CHAPTER 6

Allen, J. F., and W. Martin. 2007. "Evolutionary Biology: Out of Thin Air." *Nature* 445:610–612.

Anbar, A., et al. 2007. "A Whiff of Oxygen Before the Great Oxidation Event?" *Science* 317:1903–1906.

Diamond, J. 2005. *Collapse: How Societies Choose to Fail or Succeed.* New York: Penguin.

Barcroft, J. 1914. *The Respiratory Function of the Blood*. Cambridge: Cambridge University Press.

Battistuzzi, F. U., et al. 2004. "A Genomic Timescale of Prokaryote Evolution: Insights Into the Origin of Methanogenesis, Phototrophy, and the Colonization of Land." *BMC Evolutionary Biology* 4: 44.

Collins, S. 1996. "The Limit of Human Adaptation to Starvation." *Nature Medicine* 1:810–814.

Daly, H. E., and K. N. Townsend, eds. 1993. *Valuing the Earth: Economics, Ecology, Ethics*. Cambridge, Mass.: The MIT Press.

Gerschman, R., et al. 1954. "Oxygen Poisoning and X-Irradiation: A Mechanism in Common." *Science* 119:623–626.

Gorby, Y. A., et al. 2006. "Electrically Conductive Bacterial Nanowires Produced by *Shewanella oneidensis* Strain MR-1 and Other Microorganisms." *Proceedings of the National Academy of Sciences of the United States of America* 103:11358–11363.

Griffin, B. M., et al. 2007. "Nitrite: An Electron Donor for Anoxygenic Photosynthesis." *Science* 316:1870.

Grocott, M. P. W., et al. 2009. "Arterial Blood Gases and Oxygen Content in Climbers on Mount Everest." *New England Journal of Medicine* 360:240–249.

Kasting, J. F. 2006. "Earth Sciences: Ups and Downs of Ancient Oxygen." *Nature* 443:643–645.

Lane, N. 2003. *Oxygen: The Molecule That Made the World*. Oxford: Oxford University Press.

Office of Polar Programs. 2009. National Science Foundation Amundsen-Scott South Pole Station. http://www.nsf.gov/od/opp/support/southp.jsp.

Pianka, E. R. 1994. *Evolutionary Ecology*. 5th ed. New York: Harper-Collins.

Rao, A., et al. 2004. "Efficiency of Electrochemical Systems." *Journal of Power Sources* 134:181–184.

Smith, S. M., et al. 2001. "Nutritional Status Assessment in Semiclosed Environments: Ground-Based and Space Flight Studies in Humans." *Journal of Nutrition* 131:2053–2061.

Swenson, R. 1997. "Autocatakinetics, Evolution, and the Law of Maximum Entropy Production: A Principled Foundation Towards the Study of Human Ecology." In *Advances in Human Ecology*, ed. L. Freese, 6:1–47. Greenwich, Conn.: JAI.

Tainter, J. A. 1988. *The Collapse of Complex Societies*. Cambridge: Cambridge University Press.

Wrighton, M. S., et al. 1975. "Photoassisted Electrolysis of Water by Irradiation of a Titanium Dioxide Electrode." *Proceedings of the National Academy of Sciences of the United States of America* 72:1518–1522.

CHAPTER 7

Bischoff-Ferrari, H. A., et al. 2004. "Effect of Vitamin D on Falls: A Meta-Analysis." *Journal of the American Medical Association* 291:1999–2006.

Bushinsky, D. A. 2001. "Acid-Base Imbalance and the Skeleton." *European Journal of Nutrition* 40 (5): 238–244.

Daly, H. E., and K. N. Townsend, eds. 1993. *Valuing the Earth: Economics, Ecology, Ethics.* Cambridge, Mass.: The MIT Press.

Drysdale, A. E., et al. 2004. "The Minimal Cost of Life in Space. *Advances in Space Research* 34 (7): 1502–1508.

Holick, M. F. 2004. "Vitamin D: Importance in the Prevention of Cancers, Type 1 Diabetes, Heart Disease, and Osteoporosis." *American Journal of Clinical Nutrition* 79:362–371.

Holick, M. F. 2006. "Resurrection of Vitamin D Deficiency and Rickets." *Journal of Clinical Investigation* 116:2062–2067.

Holick, M. F. 2007. "Vitamin D Deficiency." *New England Journal of Medicine.* 357:266–281.

Jones, N. L. 1997. *Clinical Exercise Testing.* 4th ed. Philadelphia: W. B. Saunders.

Jones, S. 2006. "New Electricigens Get Wired." *National Review of Microbiology* 4:642–643.

Leya, M., et al. 2005. "Thermodynamic Efficiency of an Intermediate Band Photovoltaic Cell with Low-Threshold Auger Generation." *Journal of Applied Physics* 98 (4): article 044905, August 15.

McGilloway, R. L., and R. W. Weaver. 2004. "Effects of Drying on Nitrification Activity in Zeoponic Medium Used for Long-Term Space Missions." *Habitation* 10:15–19.

Mumpton, F. A. 1999. "La Roca Magica: Uses of Natural Zeolites in Agriculture and Industry." *Proceedings of the National Academy of Sciences of the United States* 96:3463–3470.

Pianka, E. R. 1994. *Evolutionary Ecology.* 5th ed. New York: Harper-Collins.

Planel, H. 2004. *Space and Life: An Introduction to Space Biology and Medicine.* Boca Raton, Fla.: CRC.

Rettberg, P., et al. 1988. "Biological Dosimetry to Determine the UV Radiation Climate Inside the MIR Station and Its Role in Vitamin D Biosynthesis." *Advances in Space Research* 22 (12): 1643–1652.

Rieder, R., et al. 1997. "The Chemical Composition of Martian Soil and Rocks Returned by the Mobile Alpha Proton X-ray Spectrometer: Preliminary Results from the X-ray Mode." *Science* 278:1771–1774.

Shearer, M. J. 1995. "Vitamin K." *Lancet* 345:229–234.

Smith, S. M., et al. 2001. "Nutritional Status Assessment in Semiclosed Environments: Ground-Based and Space Flight Studies in Humans." *Journal of Nutrition* 131: 2053–2061.

Yamashita, M., et al. 2006. "An Overview of Challenges in Modeling Heat and Mass Transfer for Living on Mars." *Annals of the New York Academy of Sciences* 1077: 232–243.

Zhu, H., and R. J. Kee. 2006. "Thermodynamics of SOFC Efficiency and Fuel Utilization as Functions of Fuel Mixtures and Operating Conditions." *Journal of Power Sources* 161 (2): 957–964.

CHAPTER 8

Buckey, J. C. 2006. *Space Physiology*. New York: Oxford University Press.

Burger, A. G. 2004. "Environment and Thyroid Function." *Journal of Clinical Endocrinology & Metabolism* 89:1526–1528.

Christiansen, P. 2002. "Mass Allometry of the Appendicular Skeleton in Terrestrial Mammals." *Journal of Morphology* 251:195–209.

Cogoli, A., et al. 1984. "Cell Sensitivity to Gravity." *Science* 225:228.

Convertino, V. A. 1991. "Neuromuscular Aspects in the Development of Exercise Countermeasures." *The Physiologist* 34:S125–S128.

Do, N. V., et al. 2004. "Elevation in Serum Thyroglobulin During Prolonged Antarctic Residence: Effect of Thyroxine Supplement in the Polar 3,5,3'-Triiodothyronine Syndrome." *Journal of Clinical Endocrinology & Metabolism* 89:1529–1533.

Fitts, R. H, et al. 2000. "Physiology of a Microgravity Environment. Invited Review: Microgravity and Skeletal Muscle." *Journal of Applied Physiology* 89:823–839.

Friedman, E. M., and D. A. Lawrence. 2002. "Environmental Stress Mediates Changes in Neuroimmunological Interactions." *Toxicological Sciences* 67:4–10.

Greenleaf, J. E. 2004. *Deconditioning and Reconditioning*. Boca Raton, Fla.: CRC.

Guyton, A. C., and J. E. Hall, eds. 2000. *Textbook of Medical Physiology*. Philadelphia: W. B. Saunders.

Hain, T. C., et al. 1999. "Mal de Debarquement." *Archives of Otolaryngology—Head and Neck Surgery* 125 (6): 615–620.

Jenkins, S., et al. 2011. "Endocrine-Active Chemicals in Mammary Cancer Causation and Prevention." *Journal of Steroid Biochemistry and Molecular Biology*. PMID: 21729753.

Kalpana, V., J. L. Thompson, and D. A. Riley. 1998. "Sarcomere Lesion Damage Occurs Mainly in Slow Fibers of Reloaded Rat Adductor Longus Muscles." *Journal of Applied Physiology* 85 (3): 1017–1023.

Kim, N. W., and V. M. Sanders. 2006. "It Takes Nerve to Tell T and B Cells What to Do." *Journal of Leukocyte Biology* 79:1093–1094.

Lang, T., et al. 2004. "Cortical and Trabecular Bone Mineral Loss from the Spine and Hip in Long-Duration Space Flight." *Journal of Bone Mineral Research*. doi: 10.1359/jbmr.040307.

Marcu, O., et al. 2011. "Innate Immune Responses of *Drosophila melanogaster* Are Altered by Spaceflight." *PLoS One* 6 (1): e15361 (January 11).

Moynihan, J. A., and F. M. Santiago. 2007. "Brain Behavior and Immunity: Twenty Years of T Cells." *Brain, Behavior, and Immunity* 21:872–880.

NASA. 2002. *Neurolab Spacelab Mission: Neuroscience Research in Space Results from the STS-90 Neurolab Spacelab Mission.* Ed. J. C. Buckey Jr. and J. L. Homick. NASA SP-2003-535. Houston, Tex.: Lyndon. B. Johnson Space Center.

National Osteoporosis Foundation (NOF). 1998. "Osteoporosis: Prevention, Diagnosis, and Treatment." *Osteoporosis International* 8 (suppl. 4): S1–S88.

Nichols, H. L., et al. 2006. "Proteomics and Genomics of Microgravity." *Physiological Genomics* 26:163–171.

Nikawa, T., et al. 2004. "Skeletal Muscle Gene Expression in Space-Flown Rats." *FASEB Journal* 18:522–524.

Parmet, A. J., and K. K. Gillingham. 2002. "Spatial Orientation." In *Fundamentals of Aerospace Medicine*, 3rd ed., ed. R. L. DeHart and J. L. Davis. 3rd ed., 184–244. Philadelphia: Lippincott, Williams & Wilkins.

Perhonen, M. A., et al. 2001. "Cardiac Atrophy After Bed Rest and Spaceflight." *Journal of Applied Physiology* 91 (2): 645–653.

Raisz, L. G. 2005. "Pathogenesis of Osteoporosis: Concepts, Conflicts, and Prospects." *Journal of Clinical Investigation* 115 (12): 3318–3325.

Rosen, C. J. 2003. "Restoring Aging Bones." *Scientific American* 288:70–77.

Schmidt-Nielsen, K. 1984. *Scaling: Why Is Animal Size So Important?* Cambridge: Cambridge University Press.

Sonnenfeld, G., and W. T. Shearer. 2002. "Immune Function During Spaceflight." *Nutrition* 18:899–903.

Stein, T. P., and C. E. Wade. 2003. "Protein Turnover in Atrophying Muscle: From Nutritional Intervention to Microarray Expression Analysis." *Current Opinion in Clinical Nutrition & Metabolic Care* 6:95–102.

Tracey, K. J. 2005. "Fat Meets the Cholinergic Anti-Inflammatory Pathway." *Journal of Experimental Medicine* 202:1017–1021.

Trappe, S., et al. 2009. "Exercise in Space: Human Skeletal Muscle After Six Months Aboard the International Space Station." *Journal of Applied Physiology* 106: 1159–1168.

Turner, R. T. 2000. "Physiology of a Microgravity Environment: What Do We Know About the Effects of Spaceflight on Bone?" *Journal of Applied Physiology* 89:840–847.

Vico, L., et al. 2000. "Effects of Long-Term Microgravity Exposure on Cancellous and Cortical Weight-Bearing Bones of Cosmonauts." *The Lancet* 355:1607–1611.

Vogel, S. 2003. "Size and Scale." Chapter 3 of *Comparative Biomechanics: Life's Physical World.* Princeton, N.J.: Princeton University Press.

WHO Study Group. 1994. Assessment of Fracture Risk and Its Application to Screening for Postmenopausal Osteoporosis. WHO Tech Rep Ser 843. Geneva: WHO.

CHAPTER 9

Amsel J., et al. 1982. "Relationship of Site-Specific Cancer Mortality Rates to Altitude." *Carcinogenesis* 3:461–465.

Baker, D. N., et al. 2004. "An Extreme Distortion of the Van Allen Belt Arising from the 'Halloween' Solar Storm in 2003." *Nature* 432:878–881.

Daly, M. J., et al. 2007. "Protein Oxidation Implicated as the Primary Determinant of Bacterial Radioresistance." *PLoS Biology* 5 (4).

Cucinotta, F. A., et al. 2004. "Uncertainties in Estimates of the Risks of Late Effects from Space Radiation." *Advances in Space Research.* 34:1383–1389.

Cucinotta, F. A., et al. 2005. Managing Lunar and Mars Mission Radiation Risks. Part 1: Cancer Risks, Uncertainties, and Shielding Effectiveness. NASA/TP-2005-213164.

Durante M., and F. A. Cucinotta. 2008. "Heavy Ion Carcinogenesis and Human Space Exploration." *Nature Reviews Cancer* 8:465–472.

Fornace, A. J., et al. 2000. "Radiation Therapy." In *The Molecular Basis of Cancer,* 2nd ed., ed. J. Mendelsohn et al. Philadelphia: Saunders.

Fuglesang, C., et al. 2006. "Phosphenes in Low Earth Orbit: Survey Responses from 59 Astronauts." *Aviation, Space, & Environmental Medicine* 77:449–452.

Kiefer, J., et al. 1994. "Mutation Induction by Heavy Ions." *Advances in Space Research* 14:257–265.

Lobrich, M., et al. 1995. "Repair of X-Ray Induced DNA Double Strand Breaks in Specific Noti Restriction Fragments in Human Fibroblasts: Joining of Correct and Incorrect Ends." *Proceedings of the National Academy of Sciences of the United States* 92:12050–12054.

NASA. 1999. Special Publication, Human Exploration of Mars: The Reference Mission of the NASA Mars Exploration Study Team. NASA-SP 6107.

Preston, D. L., et al. 2003. "Studies of Mortality of Atomic Bomb Survivors. Report 13: Solid Cancer and Noncancer Disease Mortality: 1950–1997." *Radiation Research* 160:381–407.

Schull, W. J. 1998. "The Somatic Effects of Exposure to Atomic Radiation: The Japanese Experience, 1947–1997." *Proceedings of the National Academy of Sciences of the United States* 95 (10): 5437–5441.

Simon, S. L., et al. 2006. "Fallout from Nuclear Weapons Tests and Cancer Risk." *American Scientist* 94:48–57.

Townsend, L. W. 2001. "Radiation Exposures of Aircrew in High-Altitude Flight." *Journal of Radiological Protection* 21.

Wan, X. S., et al. 2006. "Protection Against Radiation-Induced Oxidative Stress in Cultured Human Epithelial Cells by Treatment with Antioxidant Agents." *International Journal of Radiation Oncology * Biology * Physics* 64 (5): 1475–1481.

CHAPTER 10

Berson, D. M., et al. 2002. "Phototransduction by Retinal Ganglion Cells That Set the Circadian Clock." *Science* 295:1070–1082.

Bray, M. S., et al. 2008. "Disruption of the Circadian Clock Within the Cardiomyocyte Influences Myocardial Contractile Function, Metabolism, and Gene Expression." *American Journal of Physiology—Heart and Circulatory Physiology* 294: H1036–H1047.

Charles, J. B. 1999. "Human Health and Performance Aspects of Mars Design Reference Mission of July 1997." In *Proceedings: First Biennial Space Biomedical Investigators' Workshop*, 80–93. League City, Texas.

Clement, G. 2003. "The Musculoskeletal System in Space." Chapter 5 of *Fundamentals of Space Medicine*, 173–204. Dordrecht: Klewer.

Drake, Bret G., ed. 1998. Reference Mission Version 3.0 Addendum to the Human Exploration of Mars: The Reference Mission of the NASA Mars Exploration Study Team. NASA, Johnson Space Center.

La Duc, M. T., et al. 2004. "Microbial Monitoring of Spacecraft and Associated Environments." *Microbial Ecology* 47:150–158.

Merrow, M., et al. 2006. "The Right Place at the Right Time: Regulation of Daily Timing by Phosphorylation." *Genes and Development* 20 (19): 2629–2633.

NASA. 1999. *Human Exploration of Mars: The Reference Mission of the NASA Mars Exploration Study Team*. NASA-SP 6107.

NRC Steering Group for the Workshop on Biology-Based Technology for Enhanced Space Exploration (NRC). 1998. *Report on the Workshop on Biology-Based Technology to Enhance Human Well-Being in Extended Space Exploration*. Washington, D.C.: National Research Council.

Roman, M. C., and P. O. Wieland. 2005. Microbiological Characterization and Concerns of the International Space Station Internal Active Thermal Control System, 11–14 July 2005. International Conference on Environmental Systems, Rome, Italy. SAE-05ICES-193. NASA Marshall Space Flight Center.

Sridhar, K. R., et al. 2000. "*In-situ* Resource Utilization Technologies for Mars Life Support Systems." *Advanced Space Research* 25:249–255.

Yamaguchi, S., et al. 2003. "Synchronization of Cellular Clocks in the Suprachiasmatic Nucleus." *Science* 302:1408–1412.

Wilson, J. W., et al. 2007. "Space Flight Alters Bacterial Gene Expression and Virulence and Reveals a Role for Global Regulator Hfq." *Proceedings of the National Academy of Sciences of the United States* 104 (41): 16299–16304.

Young, M. E. 2006. "The Circadian Clock Within the Heart: Potential Influence on Myocardial Gene Expression, Metabolism, and Function." *American Journal of Physiology—Heart and Circulatory Physiology* 290 (1): H1–16.

CHAPTER 11

Bandfield, J. L. 2007. "High-Resolution Subsurface Water-Ice Distributions on Mars." *Nature* 447:64–67.

Berkovich, Y. A., et al. 2004. "Evaluating and Optimizing Horticultural Regimes in Space Plant Growth Facilities." *Advanced Space Research* 34 (7): 1612–1618.

Byrne, S., and A. P. Ingersoll. 2003. "A Sublimation Model for Martian South Polar Ice Features." *Science* 299:1051–1053.

Chown, M. 2010. "Set Shields to Stunning." *New Scientist* 207:39–41.

Fenton, L. K., et al. 2007. "Global Warming and Climate Forcing by Recent Albedo Changes on Mars." *Nature* 446:646–649.

Hand, E. 2008. "Mars Exploration: Phoenix: A Race Against Time." *Nature* 456: 690–695.

Harris, G. L. 2001. *The Origins and Technology of the Advanced Extravehicular Space Suit*. American Astronautical Society History Series 24. San Diego, Calif.: Univelt, Inc.

Hecht, M. H., et al. 2009. "Detection of Perchlorate and the Soluble Chemistry of Martian Soil at the Phoenix Lander Site." *Science* 325:64–67.

Hess, S. L., et al. 1979. "The Seasonal Variation of Atmospheric Pressure on Mars as Affected by the South Polar Cap." *Journal of Geophysical Research* 84:2923–2927.

Holt, J. W., et al. 2008. "Radar Sounding Evidence for Buried Glaciers in the Southern Mid-Latitudes of Mars." *Science* 322:1235–1238.

Hublitz, I., et al. 2004. "Engineering Concepts for Inflatable Mars Surface Greenhouses." *Advanced Space Research* 34 (7): 1546–1551.

Leovy, C. 2001. "Weather and Climate on Mars." *Nature* 412:245–249.

Mader, T. H., et al. 2011. "Optic Disc Edema, Globe Flattening, Choroidal Folds, and Hyperopic Shifts Observed in Astronauts After Long-Duration Space Flight." *Ophthalmology* 118:2058–2069.

Malin, M. C., et al. 2001. "Observational Evidence for an Active Surface Reservoir of Solid Carbon Dioxide on Mars." *Science* 294:2146–2148.

Moore, J. M. 2004. "Mars: Blueberry Fields for Ever." *Nature* 428:711–712.

Okubo, C. H., and A. S. McEwen. 2007. "Fracture-Controlled Paleo-Fluid Flow in Candor Chasma, Mars." *Science* 315:983–985.

Phillips, R. J., et al. 2011. "Massive CO_2 Ice Deposits Sequestered in the South Polar Layered Deposits of Mars." *Science* 332:838–841.

Rapp, D. 2006. "Radiation Effects and Shielding Requirements in Human Missions to the Moon and Mars." *Mars* 2:46–71.

Smith, P., et al. 2009. "H_2O at the Phoenix Landing Site." *Science* 325:58–61.

Whiteway, J. A., et al. 2009. "Mars Water-Ice Clouds and Precipitation." *Science* 325:68–70.

Yamashita, M., et al. 2006. "An Overview of Challenges in Modeling Heat and Mass Transfer for Living on Mars." *Annals of the New York Academy of Sciences* 1077:232–243.

Yen, A. S., et al. 2005. "An Integrated View of the Chemistry and Mineralogy of Martian Soils." *Nature* 436:49–54.

Zubrin, R., with R. Wagner. 1996. *The Case for Mars: The Plan to Settle the Red Planet and Why We Must.* New York: Touchstone.

CHAPTER 12

Ádámkovics, M., et al. 2007. "Widespread Morning Drizzle on Titan." *Science* (October 11). doi:10.1126/science.1146244.

Bird, M. K., et al. 2005. "The Vertical Profile of Winds on Titan." *Nature* 438:800–802.

Bolton, S. J, et al. 2002. "Ultrarelativistic Electrons in Jupiter's Radiation Belts." *Nature* 415:987–991.

Brown, M. 2010. *How I Killed Pluto and Why It Had It Coming.* New York: Spiegel & Grau.

Choueiri, E. Y. 2009. "New Dawn for Electric Rockets." *Scientific American* 300 (2): 58–65.

Griffin, M. D., and J. R. French. 2004. "Thermal Control." Chapter 9 of *Space Vehicle Design*, 2nd ed. AIAA Education Book Series. Reston, Va.: American Institute of Astronautics and Engineering.

Lancaster, N. 2006. "Linear Dunes on Titan." *Science* 312:702–703.

McCord, T. B, et al. 2006. "Ceres, Vesta, and Pallas: Protoplanets, Not Asteroids." *EOS, Transactions, American Geophysical Union* 87:10.

Miyakawa, S., et al. 2002. "The Cold Origin of Life: B. Implications Based on Pyrimidines and Purines Produced from Frozen Ammonium Cyanide Solutions." *Origins of Life and Evolution of Biospheres* 32 (3): 209–218.

Owen, T. 2005. "Planetary Science: Huygens Rediscovers Titan." *Nature* 438:756–758.

Service, R. F. 2006. "Nanocolumns Give YBCO Wires a Big Boost." *Science* 311: 1850–1851.

Stanley, S., and J. Bloxham. 2004. "Convective-Region Geometry as the Cause of Uranus' and Neptune's Unusual Magnetic Fields." *Nature* 428:151–153.

Stewart, I. 2006. "Ride the Celestial Subway." *New Scientist* (March 25–31): 32–36.

Thomas, P. C., et al. 2005. "Differentiation of the Asteroid Ceres as Revealed by Its Shape." *Nature* 437:224–226.

Tobias, O. 2005. "Planetary Science: Huygens Rediscovers Titan." *Nature* 438:756–757.

Troutman, P. A. (NASA Langley Research Center) et al. 2003. Revolutionary Concepts for Human Outer Planet Exploration (HOPE). http://nasa-academy.org/soffen/travelgrant/bethke.pdf.

Uijlenhoet, R., et al. 2003. "Variability of Raindrop Size Distributions in a Squall Line and Implications for Radar Rainfall Estimation." *Journal of Hydrometeorology* 4:43–61.

Walsh, K. J. 2009. "Asteroids: When Planets Migrate." *Nature* 457:1091–1093.

Wirth, B. D. 2007. "How Does Radiation Damage Materials?" *Science* 318:923–924.

CHAPTER 13

Benedict, G. F., et al. 2006. "The Extrasolar Planet Epsilon Eridani B: Orbit and Mass." *Astronomical Journal* 132:2206–2218.

Blackstone, E., et al. 2005. "H_2S Induces a Suspended Animation–Like State in Mice." *Science* 308:518.

Briskie, J. V., and M. Mackintosh. 2004. "Hatching Failure Increases with Severity of Population Bottlenecks in Birds." *Proceedings of the National Academy of Sciences of the United States* 101 (2): 558–561.

Brogi, M., et al. 2009. "Dynamical Stability of the Inner Belt Around Epsilon Eridani." *Astronomy and Astrophysics* 499 (2): L13–L16.

Bussard, R. W. 1960. "Galactic Matter and Interstellar Flight." *Astronautica Acta* 6:179.

Butler, P., et al. 2004. "A Neptune-Mass Planet Orbiting the Nearby M Dwarf GJ 436." *Astrophysical Journal* 617:580–588.

Carey, H. V., et al. 2003. "Mammalian Hibernation: Cellular and Molecular Responses to Depressed Metabolism and Low Temperature." *Physiological Reviews* 83:1153–1181.

Cassan, A., et al. 2012. "One or More Bound Planets per Milky Way Star from Microlensing Observations." *Nature* 481 (7380): 167–169.

Chown, S. L., et al. 2007. "Scaling of Insect Metabolic Rate Is Inconsistent with the Nutrient Supply Network Model." *Functional Ecology* 21:282–290.

Colman, R. J., et al. 2009. "Caloric Restriction Delays Disease Onset and Mortality in Rhesus Monkeys." *Science* 325:201–204.

Croswell, K. 1996. *The Alchemy of the Heavens: Searching for Meaning in the Milky Way*. New York: Anchor.

Danzl, D. F., and R. S. Pozos. 1994. "Current Concepts: Accidental Hypothermia." *New England Journal of Medicine* 331 (26): 1756–1760.

Diehl, R., et al. 2006. "Radioactive [26]Al from Massive Stars in the Galaxy." *Nature* 439:45–47.

Dobson, G. P., and J. P. Headrick. 1995. "Bioenergetic Scaling: Metabolic Design and Body-Size Constraints in Mammals." *Proceedings of the National Academy of Sciences of the United States* 92:7317–7321.

Fressin, F., et al. 2011. "Two Earth-Sized Planets Orbiting Kepler-20." *Nature* (December 20).

Gai, M. 2003. "Open Questions in Stellar Helium Burning Studied with Real Photons." In *Fission and Properties of Neutron-Rich Nuclei*, ed. J. H. Hamilton et al. Hackensack, N.J.: World Scientific Publishing.

Garlick, M. A. 2002. *The Story of the Solar System*. New York: Cambridge University Press.

Gaudi, B. S., et al. 2008. "Discovery of a Jupiter/Saturn Analog with Gravitational Microlensing." *Science* 319:927–930.

Gredilla R., and G. Barja. 2005. "Minireview: The Role of Oxidative Stress in Relation to Caloric Restriction and Longevity." *Endocrinology* 146 (9): 3713–3717.

Hogue, C. W., et al. 2006. "Cardiopulmonary Bypass Management and Neurologic Outcomes: An Evidence-Based Appraisal of Current Practices." *Anesthesia & Analgesia* 103:21–37.

Kalas, P., et al. 2008. "Optical Images of an Exosolar Planet Twenty-five Light-Years from Earth." *Science* 322:1345–1348.

Pallé, E., et al. 2009. "Earth's Transmission Spectrum from Lunar Eclipse Observations." *Nature* 459:814–816.

Petersen, J. M., et al. 2011. "Hydrogen Is an Energy Source for Hydrothermal Vent Symbioses." *Nature* 476:176–180.

Piantadosi, C. A. 2003. *The Biology of Human Survival—Life and Death in Extreme Environments.* New York: Oxford University Press.

Pierre Auger Collaboration, et al. 2007. "Correlation of the Highest-Energy Cosmic Rays with Nearby Extragalactic Objects." *Science* 318:938–943.

Sagan, C., et al. 1993. "A Search for Life on Earth from the Galileo Spacecraft." *Nature* 365:715–718.

Sasselov, D. D. 2008. "Extrasolar Planets." *Nature* 451:29–31.

Saumon, D., et al. 1996. "A Theory of Extrasolar Giant Planets." *Astrophysical Journal* 460:993–1018.

Schneider, J., et al. 2011. "Defining and Cataloging Exoplanets: The Exoplanet.eu Database." *Astronomy & Astrophysics* A79. http://dx.doi.org/10.1051/0004-6361 201116713; http://exoplanet.eu.

Tinetti, G., et al. 2007. "Water Vapor in the Atmosphere of a Transiting Extrasolar Planet." *Nature* 448:169–171.

INDEX

acceleration, 30, 65, 67, 82, 83, 182, 154, 155, 174, 240, 243; centripetal, 24
acclimation, 31
acclimatization, 30, 31, 65, 216
accommodation, 32
adaptagent, 31
adaptation, 30–36, 39, 109, 126, 156, 161, 169
adenosine diphosphate (ADP), 130, 144
adenosine triphosphate (ATP), 130, 131, 144, 146, 159, 160
Aeromedical Laboratory, 62, 63, 65
aerospace medicine, 61, 62, 65, 66, 67
air, 5, 7–8, 28–29, 32–33, 57, 62, 64, 68–69, 71, 77, 82–83, 86, 94, 101, 105–6, 125, 127, 129, 130–31, 141, 193, 196, 222, 232, 243; heavier-than, 13, 152; conditioning, 39; temperatures, 208

Air Force, 63–67, 81
Air Service Medical Research Laboratory, 61, 62
ALARA, 169, 172, 175, 231
albedo, 97, 211, 239
Aldrin, E. E., 72, 216
allele, 34, 39
allometry, 153
Alpha Centauri, 237–39, 249
Alpha Magnetic Spectrometer, 17, 79
Altiplano, Andean, 33
altitude, 29, 32–35, 58, 62–69, 75, 84, 87, 105–6, 130, 171–73, 175, 195, 207, 209; high-, 64
aluminum, 62, 77, 92, 97, 99, 103, 129, 148, 183, 184, 212
Amundsen, R., 48, 49, 54, 90, 124; Gulf, 51
Andes, 33
Andromeda, 44, 249

Antarctica, 5, 7, 48, 51, 55, 90, 124, 167, 168, 204, 207, 208, 211, 220, 224, 228
antigravity, 63, 136, 159, 160, 162, 164
antimatter, 17, 236, 244
antioxidant, 131, 133, 135, 151, 181, 187
Apollo, 1, 2, 7, 8, 21, 23, 41, 71–73, 80, 88, 93–95, 104, 107, 165, 183, 188, 205–6, 216; *Apollo 1*, 71; *Apollo 7*, 71; *Apollo 8*, 72; *Apollo 10*, 72; *Apollo 11*, 13, 40, 72, 74, 83; *Apollo 12*, 101; *Apollo 13*, 72, 148; *Apollo 15*, 73; *Apollo 16*, 147; *Apollo 17*, 72, 94, 147; Project, xii, 70–71, 171; Apollo-Soyuz, 73
Arctic, 39, 48–53, 55–56, 61, 92, 209; Canadian Archipelago, 56
Armstrong line, 69, 76, 213, 214, 215
Armstrong, H. G., 63–66, 69
Armstrong, N. A., 72
Army Air Corps, 62–63
Artemia, 147
asteroids, xii, xiii, 7, 18–19, 23, 25–26, 40–41, 88–89, 99, 113, 189, 205, 221–24
asteroid belt, 22, 36, 220–23, 235
astrobiology, 6, 7, 91
astronauts, xii, xiii, 1, 2, 5, 14, 17–19, 21, 23, 37, 39–40, 46, 58, 69–76, 80–81, 83–90, 94–95, 101, 104–7, 111, 134, 136–38, 140, 145, 149, 154–60, 162–66, 172, 175, 179, 182–85, 188–91, 195, 197–98, 200, 204–5, 211–12, 215, 216, 229, 232, 247–49
astronomical unit (AU), 37–38, 41, 120, 122–23, 222–23, 229, 231, 234, 238, 240
astrophysics, 6, 24, 67, 90
atmosphere, xiv, 7, 8, 9, 25, 28–30, 36, 38–39, 41, 45, 51, 55, 57–58, 60, 67, 69, 71, 77, 81–82, 84–85, 87, 93–95, 99, 100–101, 105–6, 108, 110, 112, 120–21, 124–25, 130–33, 141–43,

145, 171, 173–75, 183, 191–92, 194–95, 205–16, 224–25, 227–31, 233, 239–43, 246
atomic power, 60, 120
auroras, 51, 54, 55, 58, 226
aviation, 5, 59, 61–67, 81, 155

B-17, 63
B-29, 63
Barents Sea, 50, 51
barometric pressure, 29, 32, 68, 69, 106, 130, 195, 207
baryon, 173, 174, 175
basal energy expenditure (BEE), 136
basal metabolic rate (BMR), 144–145
behavioral adaptation, 39, 109, 126
Berkner, L. V., 55
Bernard, C., 31
biofilm, 194, 195
biofouling, 194
biofuel, 139
biogeography, 8, 37, 237
biomass, 212, 213
biomedicine, 3, 30
Birkeland, K., 54, 55
bisphosphonates, 158
blood, 32–35, 40, 108, 129–30, 133, 143–44, 151, 157, 160, 162, 165, 178–79, 181–82, 205; clot, 30, 179; hemoglobin, 34; platelet, 183; pressure, 31, 33, 39, 199; vessels, 30–33, 35, 150, 157, 178, 181, 182; volume, 40, 162
bone, 18, 27–28, 135–36, 148–58, 168; density, 27, 136, 145, 157, 198; fragility, 158–59; loss, 27, 100, 136, 156–59, 189, 198, 204, 224; marrow, 28, 178, 180, 182; metabolism, 149; remodeling, 158; strength, 151, 153, 157, 158; stress, 154
brain, 66, 110, 129, 131, 150, 155–56, 167–68, 172, 175, 179, 182–84, 199, 205, 247–48

Bremsstrahlung, 184
Brown, M., 220

cabin, 13, 29–30, 35, 58–60, 63, 65, 84, 86, 105, 121, 145, 148–49, 189, 193, 204
calcium, 97–99, 136, 145, 148–51, 158–59, 198, 209, 212; manganese, 133
Callisto, 42, 220, 227, 229, 231, 234
cancer, 30, 36, 73, 101, 150–51, 158, 164, 166–68, 171–72, 175, 177–78, 180–83, 185–87, 197
Cannon, W., 31
capillaries, 33, 35, 160, 161, 164
carbon (C), 77, 126, 132, 135, 143, 147, 177, 193, 207, 225, 242, 244–45, 247; cycle, 134–35, 191, 193
carbon dioxide (CO_2), 29, 137, 211
carbon monoxide (CO), 98, 207, 244
carbonaceous chondrites, 223–24
Carlson, L. D., 63
Cassini-Huygens, 19, 20
Ceres, 19, 36, 42, 100, 122–23, 219–20, 223–25, 229
Cernan, E. A., 14, 72, 94
cetaceans, 129, 154
Chaffee, R. B., 71
Challenger, 69, 75–77, 85
Chandrayaan-1, 98
Chernobyl, 120, 181–82
China, 14, 91–92, 103, 205, 218
China National Space Administration (CNSA), 91
chlorofluorocarbons (CFCs), 5
chlorophyll, 132, 135
chloroplast, 132, 135
chromosome, 34, 147, 177, 178, 181
chronobiology, 199–200
circadian rhythm, 199–200
Clementine, 96
climate change, 5, 41, 56, 92, 110, 118, 211, 215

cold, xiv, 13, 30, 31–32, 36, 38–39, 42, 48, 52, 59, 63, 66, 95–97, 102, 109, 112, 118–19, 123–25, 141, 148, 167–68, 193, 207–8, 216, 221, 224, 227–32, 234, 237, 243, 247; trap, 90, 96; water immersion, 66
Cold War, 4, 45, 49, 70, 78, 88
Collins, M., 72
Columbia, 69, 72, 75–78, 87, 107, 111, 138
Columbia Accident Investigation Board (CAIB), 15, 76
Compton effect, 176
confidence interval, 186
Constellation, Project, xii, xiii, 1, 14, 16, 22, 23, 78, 79, 239
coronal mass ejection (CME), 95
cosmonaut, 70, 72, 74, 81, 83–84, 86–87, 137
countermeasures, 28, 156–57, 159, 165, 175, 185–86, 198, 213, 224
crater, 42, 90, 95–98, 102, 108, 138, 208, 222; Endeavor, 19; Shackleton, 89–90, 97
cryogenics, 234
Curiosity, 120, 193, 214, 217
cytokines, 181

Dalton's law, 32
Dawn spacecraft, 224, 225, 229, 231
decompression, 63, 105–6, 179; sickness (bends), 105, 179
deconditioning, 159, 161
Deep Space 1, 123
Deimos, 205
Deinococcus radiodurans, 187, 192
deoxyribonucleic acid (DNA), 34, 172, 177, 178, 187
desert, 31, 133, 207
Dill, D. B., 63
Discovery, 78, 79
diuresis, 39, 40, 162
Drake, F., 239, 240

dry ice, 208, 209, 210, 229
dust, 47, 91, 95, 100–101, 106, 112, 120, 192, 206–8, 211, 214, 240, 243, 244
dwarf planet, 20–21, 219–20, 223
dysautonomia, 39

Earth, xiii, xiv, xv, 5–8, 13, 16, 18–21, 24–28, 35–45, 47, 49–50, 54–59, 69–72, 74, 79–85, 87, 89–103, 107–12, 117, 119, 121–23, 125–29, 131–36, 138–39, 141, 143, 147, 148, 151–54, 156, 157, 168, 171, 173–75, 182, 183, 190–93, 196, 198, 200, 205–12, 215, 217, 220–27, 229–32, 237, 241, 242, 250; atmosphere, 9, 29, 195, 215; biosphere, 142; ecosystem, 143; gravity, 17, 24, 154–55, 191; to LEO, 15, 23, 25, 154; super-, 38
ebullism, 69, 76, 84, 213
ecological footprint, 125–26
ecology, 37, 125–26
ecosystems, 27–28, 125, 141, 143, 192
electricigens, 139
electrolysis, 60, 127, 141, 143, 196, 232, 238
electron, 57, 58, 95, 131–33, 139, 144–45, 173–74, 176–77, 184, 236, 242
Ellesmere Island, 51–53
endocrine disruptors, 167
energy, 27–28, 42, 100, 102, 108–9, 113, 119-21, 123, 125–29, 131–49, 159, 173–76, 183–86, 209, 213, 231, 233, 236, 242, 244, 246–49; chemical, 130, 144; Department of, 21; high, 95, 144, 147, 174, 176, 178, 183–84, 211–12, 249; low, 174, 176, 184, 212; solar, 100
entropy, 109, 120
EPAS1, 34, 35
epigenetics, 36, 167; mark, 177, 178
Epsilon Eridani, 239–40, 249
Eris, 42, 122, 220, 237

escape velocity, 93, 152–53, 205, 224, 229
estivation, 246–47
Europa, 226–28
eutectic mixture, 227
evolution, xiv, 3, 110, 125–26, 131–33, 153, 243–44, 250
exoplanet, 6, 25, 38, 43, 240–43, 246
extravehicular activity (EVA), 22, 74, 105, 113, 127, 145, 156, 179, 214
extremophiles, 6, 193, 207

flight surgeons, 61–62, 64
food, 7, 9, 27, 28, 43, 48, 53–54, 60–61, 64, 121, 126, 133–38, 140–44, 146–49, 191, 206, 211, 213, 237, 247; freeze-dried, 59; preservation, 5
Fort Conger, 53
fractures, 150–51, 15–59, 198–99
free radical, 131, 172, 177, 187
Fresnel lens, 123
fusion, 22, 101–2, 235–36, 243–45

Gagarin, Y., 13, 70
Gagge, A. P., 63, 65
Galilean moons, 225–27
Ganymede, 227–29
gas, 29, 32, 42, 59–60, 69–70, 75–76, 94, 104–5, 127, 130, 137, 143, 179, 195–96, 213, 220, 225, 239–42; argon (^{40}Ar, ^{36}Ar), 29, 94, 207; carbon monoxide (CO), 207, 244; carbon dioxide (CO_2), 5, 29, 32, 94, 127, 137, 144, 195, 207, 209, 211, 242–44; embolization, 179; gangrene, 193; helium (^{4}He), 94, 243; hydrogen (H_2), 94, 127, 243; nitrogen (N_2), 29, 60, 105, 127, 196, 207, 242; methane (CH_4), 5, 94, 243, 207, 244; neon (^{20}Ne, ^{22}Ne), 94; natural, 231; oxygen (O_2), 32, 29, 94, 127, 131–32, 144, 195–96, 207, 241–42; sewer, 248
gas giant, 225, 240–41
Gemini, Project, 70–71

gene, 33–35, 39, 132, 150, 164–67, 177, 181, 194–95, 199; clock, 199; oncogenes, 181
genetics, 3, 6, 32–34, 36, 39, 66, 144, 171–74, 177, 181, 187, 191, 237, 246–47
geotropism, 146, 153
Gernhardt, M., 105
Glenn, J. H., 70
Gliese 581 (Gl581), 38, 43, 239
global positioning system (GPS), 59, 81
glucose, 35, 130, 131, 144, 145, 159, 245
Goddard, R., 81, 82
gravity, xiv, 17–18, 29, 39, 42, 45, 57, 75, 81, 83, 93, 99–101, 104, 112, 125, 146, 151–55, 163–68, 191, 195, 204–5, 214, 216, 220–29, 241; artificial, 198, 204; Earth's, 24, 154
great oxidation event (GOE), 131–32, 242
Greely, A., 52–54
greenhouse, 41, 109, 148, 192, 212; effect, 38, 207, 211, 215, 229, 242; gas, 5
Greenland, 5, 51, 53 118
Grissom, V., 71
growth factor, 181

habitable zone (HZ), 38, 41, 109, 240–41
habitat, 23–26, 56, 81, 90, 95, 101, 103–9, 119, 148, 166, 191, 206, 213–14, 216–17, 228, 233–35
Hale, E. E., 80–81
Hall, F. G., 63–64
heart, 5, 30, 31, 33, 35, 40, 73, 129, 143, 157–58, 162, 170, 179, 199–200; muscle, 162; stroke volume, 161–62
heat, 20–21, 31–32, 39, 42, 69, 94–95, 102, 104, 108–9, 120–24, 127, 129–30, 133, 135, 140, 143–46, 193, 209, 213, 226–27, 230–34, 238–39, 244; balance, 233; loss, 32, 233; protective, 76; shield, 76, 189; wave, 39

heliosphere, 174, 226
helium (He), 69, 94, 101–2, 173–74, 184, 225, 231, 243–45
hemoglobin, 33–35, 130, 160
hibernation, 163, 246–47
high earth orbit (HEO), 23–24
Himalayas, 33
homeostasis, 30–31, 247
hormone, 28, 31, 166–67; steroid, 163, 167, 181; parathyroid, 150; thyroid, 167–68
Hubble space telescope (HST), 2, 111
Human Exploration Framework Team (HEFT), 23, 170–71, 204
hydrogen (H$_2$), 22–23, 94, 96, 102, 121, 127, 131–32, 142, 174, 184, 196, 225–28, 231, 238, 242–45
hydrothermal vent, 41, 132, 242
hyperbaric chamber, 104
hyperoxia, 131
hyperventilation, 31
hypobaric chamber, 62
hypothermia, 32, 246, 247
hypoxemia, 129
hypoxia, 32–35, 63, 65–66, 105, 129–30, 171, 181, 213; anemic, 130; cytotoxic, 130; hypoxic, 130; -inducible factor (HIF), 35; stagnant, 130

ilmenite, 100–102
immune system, 28, 166–68
indigenous support of life and environment (ISLE), 118–19
In situ resource utilization (ISRU), 28, 99, 118
infection, 163, 166, 168, 18–82, 195
International Council of Scientific Unions (ICSU), 49, 55
International Geophysical Year (IGY), 5, 49, 55–57
International Polar Year (IPY), 49, 51, 54–57

International Space Station (ISS), xi, xii, 1, 7, 13–18, 22–26, 40, 74, 76–79, 85–87, 90–94, 105, 107–8, 111–13, 120–22, 126–27, 134–38, 140–41, 143, 145–46, 148, 154, 157, 159, 162, 170, 183, 194–96, 234
interstellar medium (ISM), 243–45, 248
inverse square law, 42, 120, 122
Io, 21, 226–27
irradiance, 89, 100, 122
island, 8, 37, 42–43, 50–55, 61, 72, 107, 109, 118, 181–82, 237, 245, 248, 250

James Webb Space Telescope (JWST), 24, 99, 111
jetliner, 21, 25, 29, 68–69
joule, 119, 176, 231
Jupiter, 21–22, 36, 42, 122–23, 153, 183, 220, 223–29, 231, 240–41

Karman line, 69, 195
Kepler spacecraft, 241
Kislingbury, F., 53–54
Korolyov, S., 73
Kuiper belt, 20, 22, 42, 219–20, 240

Lady Franklin, 52; Bay, 52–53
Lagrange point, 23–25, 99
Laika, 58
leukemia, 175, 182
life support, 22, 27, 42, 84–86, 90, 96, 98, 101, 104, 107, 121, 129, 139, 141–42, 145, 189, 204, 206, 212, 217; system, 8, 28, 60, 107–8, 119, 127–28, 139–41, 160, 194
light, 37, 54–55, 94, 97, 102, 120, 122–23, 147–48, 150, 167, 173, 175, 192, 199–200, 242; minutes, 237; seconds, 237; speed, 43, 47, 174, 236–37, 243, 248; ultraviolet, 95, 178; -years, 37, 38, 44, 47, 237–43, 246, 248, 249
linear energy transfer (LET), 176–77, 185–87
Logan, J., 170

Lovelace, W. R., 63–64
Lovell, J. A., 72
low earth orbit (LEO), xiv, 7–8, 15, 23–25, 28, 59, 126, 147, 154
Luftwaffe, 64–66; Institute for Aviation Medicine, 66
lunar: crater observation and sensing satellite, 97–98; exploration, xiii, 16, 24, 89, 90–91, 95, 98, 100, 112, 124; liquid mirror telescope (LLMT), 112; Prospector, 96–97; reconnaissance orbiter (LRO/LCROSS), 16, 97–98
lung, 29, 32–35, 101, 106, 129–33, 143–44, 175, 179, 181–82
lymphocyte, 165–66, 178–79, 182
Lyster, T. C., 61

magnesium (Mg), 97–99, 209, 212, 215, 223, 225
magnetosphere, 21, 91, 226, 229, 231; mini-, 206
mammal, 42, 43, 108, 130, 153, 177, 180, 199, 247–48
manganese (Mn), 133, 187; calcium, 133
Mars, xii, xv, 1, 2, 5–8, 14, 18–19, 22–28, 36, 38, 40–41, 69, 85, 90–93, 98–101, 104, 107–9, 113, 118–22, 125–26, 134, 138–39, 143, 146, 148, 151, 157, 175, 183, 187–97, 200, 203–20, 223–24, 229, 237, 241; atmosphere, 191, 209–11, 213, 215; dust storms, 192; escape velocity, 152–53; Express Orbiter, 208; 500 experiment, 189; gravity, 100; mission, 24, 41, 74, 78, 87, 92, 123, 137, 139, 146–50, 183, 189–90, 194–98, 203, 205, 217, 246; Moon and, 1, 2, 14, 16, 19, 22, 77–78, 88–89, 170; moons of, 205; radiation, 212; Science Laboratory, 16, 193; soils, 148, 209; spacecraft, 191–92, 195, 206; temperature, 207, 209; water on, 209–10
mass balance, 137
McClure, R., 56; McClure Strait, 51, 56

Mercury: capsule, 70; Project, 68, 70; Seven, 69
Mercury (planet), 38, 41–42, 98, 122, 153, 220, 228–29, 238, 241
metabolism, 69, 108, 139, 149, 199, 248
metals, 21, 103, 131, 133, 150–51, 177, 223–24, 234, 245
meteoroid, 222, 224; micro-, 123
methane (CH_4), 5, 94, 98, 141, 196, 207, 229–31, 242–44
microarray analysis, 164
microbe, 30, 31, 139, 187, 192–95, 218
microgravity, 8–9, 17–18, 27–28, 30, 39–40, 60, 67, 80, 82, 93, 100, 105, 113, 136, 145, 147, 151, 154–68, 191, 194–95, 204, 211
micronutrient, 135, 149–51
Milky Way, 44, 174, 238, 241, 243, 246, 248–49
minerals, 92, 135–36, 145, 148, 194, 198, 207, 209, 211–12, 224
mining, 92, 100–104, 145, 170, 213, 224, 235
Mir, 84–85, 138, 157, 160, 193
mitochondria, 135, 144, 160–65, 248
Moon, xii, xiii, 1, 2, 6–8, 13–14, 16, 18, 21–28, 40–41, 58, 70–73, 78–79, 82–83, 87–113, 118, 120–28, 148, 151, 154, 156, 170, 175, 183, 188–89, 203, 205–6, 212–14, 219, 221, 222, 224, 227, 229, 237, 242; and Mars, 1, 2, 14, 16, 19, 22, 77, 78, 88–89, 170
moonquake, 94–95
motion sickness, 26–27, 65, 156–57
Mt. Everest, 30, 69, 105, 130, 171, 195, 215
muscle, 18, 27–28, 32–33, 89, 129, 134, 145, 155, 157, 159–64, 168, 247; atrophy, 159–64, 198, 224; fatigue, 61; loss, 100, 136, 156–57, 159, 189, 198; mass, 27, 136, 144, 151, 159–62; protein, 134; skeletal, 150, 159, 162, 168; strength, 153, 161; volume, 162

musculoskeletal system, 155, 166
mutation, 34–36, 147, 172, 174, 177–78, 180, 185, 199
myofibril, 159
myoglobin, 160–61

N1 rocket, 73
NASA, xi–xiv, 2–4, 13–19, 23, 26, 28, 41, 57–59, 68, 70–80, 83–91, 96, 99, 101–8, 111–13, 120—25, 134–41, 145–48, 155, 166, 169–71, 179, 183, 185, 189, 190, 193, 195–98, 203–7, 213, 216, 221–24, 229, 236, 241
Nautilus, 60–61
near earth object (NEO), 23–4, 221–23
Neptune, 22, 36, 37, 42, 122, 153, 220, 225, 231–32, 237, 240
neuroimmune modulation (NIM), 166–68
neutron, 96, 173, 176, 184–85, 212, 244–46
New Horizons, 20–21, 96, 119–20, 231
New Zealand, 42, 237
nitric oxide (NO), 33–35, 245
nitrogen (N_2), 29, 58, 95, 132, 148, 163, 177, 191, 195, 212, 220, 231, 234
Norse, 118
North Pole, 49, 50, 52, 54, 61
Northeast Passage, 50
Northwest Passage, 48, 52, 56
nuclear reactor, 234
Nuremberg War Crimes Tribunal, 66

Oberon, 220, 231, 233–34
Oberth, H., 81–83
oncogenes, 181
orbital sciences, 15
Orion, 22–23, 111, 148, 248
orthostatic hypotension, 39–40
osteoblasts, 158
osteoclasts, 158
osteoporosis, 18, 150–51, 15–58, 197–99

Outer Space Treaty, 103
oxidation, 144, 145, 149, 245; Great
Event (GOE), 131, 242
oxygen (O₂), 9, 16, 21–23, 28–29, 58,
60–63, 65–66, 99, 109, 131–32, 137,
142, 166, 171, 177, 193, 228, 235,
244–45; enhancement ratio, 172;
partial pressure (PO₂), 130
oxygenic photosynthesis, 31, 125,
131–35, 191, 239
ozone, 95, 242; hole, 5, 56, 243; layer, 5,
148

pair production, 176
panspermia, 192, 250
papilledema, 204
parallax, 37; second, 37
parsec, 37–38, 43, 250
partial gravity, 18, 93, 99–101, 151, 156,
167–68, 205
particles, 58, 95, 101, 139, 147, 173–78,
184, 212, 226, 235, 243, 245, 249;
heavy, 175–76, 186; high-energy,
184, 211; ionizing radiation, 173, 176;
solar, 174
particulate, 101; airborne, 29
patent foramen ovale (PFO), 179
pathology, 30
Pavy, O., 53–54
Peenemünde, 58
perchlorate (ClO₄⁻), 143, 148, 209
Petermann, A., 50
Phobos, 192, 205
Phobos-Grunt spacecraft, 192, 217
Phoenix lander, 148, 209, 217
phosphenes, 184
photoelectric effect, 82, 176
photon, 132, 175–76; incident, 176
photosynthesis, 131–32, 135, 141–42,
148, 212; oxygenic, 31, 125, 131–35,
191, 239
photosystem, 132–33
physiology, 31, 46, 61, 64, 67, 162, 247;
aerospace, 64; aviation, 61, 63; cardio-

vascular, 61; European Module, 17;
limit, 30; muscle, 159
Piccard, A., 13, 63
placenta, 33, 35
planets, xiv, 1, 5–6, 14, 18–22, 25–26,
29, 35, 38–39, 41–43, 47, 57, 75, 81–
82, 87, 93, 100, 109, 125, 127, 132,
134, 139, 143, 148, 152–53, 173, 189,
191, 193, 195–97, 204–16, 222–29,
233, 236, 238–43, 246, 250; dwarf,
20–21, 219–20, 223; Red Planet, xi,
19–20, 41, 191–92, 200, 203, 218–20;
Planet X, 20
plants, 17, 31, 99, 124, 132–35, 141–42,
146–49, 153, 191, 193, 199, 212, 215
plate tectonics, 57, 81
platelet: platelet-derived growth factor
(PDGF), 181; blood, 182
Pluto, 19, 20–21, 42, 96, 98, 108, 122,
219–20, 237
plutonium, 20–21
pollutant, 5, 29, 56, 101
Polyakov, V., 85
Positron, 176, 236–37
Potocnik, H. (also Noordung, H.), 82
power, 3, 6, 21–22, 27, 43–44, 55, 60–
61, 74, 81, 90, 92, 100, 102, 104, 109,
112, 118–25, 129, 133, 138–39, 142,
144, 160, 162, 191, 199, 203, 206–7,
212, 224–25, 228, 230–35, 243–44,
248; atomic, 60, 120; horsepower,
50; nuclear, 83, 120, 124; power-to-
weight ratio, 139; solar, 20, 42, 81,
120, 122–24, 224; steam, 170; wind,
120, 207, 230
Project Paperclip, 64
proprioception, 155
protein, 33–36, 130, 132–36, 144, 151,
159–60, 163–67, 172, 177–78, 187,
195, 199
Proteus, 52–53
proton, 57–58, 85, 95, 102, 131, 144,
173–74, 184, 244–45, 249
puffy-face bird-leg syndrome, 39, 204

"Q," 43, 172, 205

radiation, 8–9, 17–18, 23, 27, 30, 42, 44, 80, 91, 94–95, 100, 106, 112, 131, 136, 147, 149–51, 166, 168–89, 192, 197, 204, 211–13, 216–17, 224, 226–34, 237, 246; belts, 57, 226, 228, 229; Cerenkov, 184; cosmic, xiv, 12, 18, 28–29, 35–36, 42–43, 57, 80, 94, 123, 139, 147–48, 169–75, 182, 186, 191, 206, 211, 213, 224, 231, 249; electro-magnetic, 173; exposure, 36, 169, 172, 175, 178, 182, 185, 197; galactic, 183; galactic cosmic (GCR), 174; gamma, 245; geomagnetic, 174; high-LET, 176, 186–87; infrared, 94; ionizing cosmic, 169, 172; ionizing particle, 173–76, 183, 187; low-energy, 176; low-LET, 176–77, 186; medical, 173; particle, 174; protection, 170–71, 181, 212; reactor, 170; sickness, 95, 182, 184, 187; solar, 41, 171, 183; solar cosmic (SCR), 174; space, 103, 173, 175, 185, 187; tolerance, 180–81; ultraviolet, 5, 175; UV-B, 150
radioisotope thermoelectric generator (RTG), 20–21, 232
rare earth element, 103
reactive oxygen species (ROS), 131, 166, 171, 172, 177
recycling, 28, 43, 98, 103, 106, 125, 128–29, 137–38, 140–43, 146–47, 196–97, 245, 249; food, 134; O_2, 35, 107, 134; waste, 121; water, 128–29, 134
red blood cells, 32, 34, 130, 151
red dwarf, 38, 238–39
regolith, 95, 97, 99–102, 104, 212, 214
relativity, 44, 155, 248–49
respiration, 64, 130, 134–35, 143–45, 191, 248
respiratory quotient (RQ), 145, 196
RNA (ribonucleic acid), 164, 178, 195, 199, 227

Rickover, H. G., 59, 170
Ride, S. K., 74
risk, xi, xiii, 1, 2, 21, 25–26, 30–31, 37, 43, 44, 46–47, 78–79, 105, 110, 118, 136, 150–51, 157–58, 164, 166, 171–72, 175, 179–89, 195, 197–99, 204, 216–17, 250
robotics, 6
rocketry, 45, 49, 57, 81
Rogers Commission, 75
rover, 73, 105, 207; Mars, 19, 148, 193, 207, 214
Russia: cosmonauts, 86; Mars mission, 246; *Phobos-Grunt*, 217; *Progress*, 87, 107; *Proton*, 85; Russian Federation, xii, 15; Space Agency, 85, 192; *Soyuz*, 85, 87, 91

Sabatier process, 141
Sagan, C., xiv, 109, 241–42
Salmonella typhimurium, 194
Salyut 1, 83, 84, 85
satellites, 5, 57, 58, 80–82, 92, 123, 126, 158, 220, 226, 232; Lunar Crater Ob-servation and Sensing, 97; SMART-1, 97; Solar and Heliospheric Observa-tory (SOHO), 99
Saturn, 19, 36, 42, 98, 122–23, 153, 220, 225–26, 229, 230–31, 241; *Cassini-Huygens*, 19
Saturn V (rocket), 23, 58, 71, 74, 84
Schley, W. S., 54
Schmidt, H., 14, 101–2
science, xi–xiv, 1–8, 15–18, 30, 42, 45–46, 49, 54–57, 70, 81, 89–91, 110, 112, 125, 135, 143, 149, 157, 186, 212, 218, 236, 246, 248, 250; International Council for (ICSU), 55; Mars Laboratory, 16, 193; National Academy of, 55, 56, 111; National Science Foundation, 3, 90; space, 2–3, 5, 79, 96; Space and Life Sciences Directorate (NASA), 189; Wars, 4

science fiction, 14, 29, 37, 47, 66, 102, 109, 117, 215, 224, 227–28, 245, 248, 250

Scott, R., 48, 49

Selene, 92, 93, 97, 227

Shackleton Crater, 89–90, 97

Shenzhou, 91

Shepard, A. B., 70

shielding, 18, 22, 28, 101, 106, 169, 170–71, 174–75, 183–85, 189, 197, 206, 213, 217, 221, 231

sievert (Sv), 176, 178, 183, 185

silver, 112, 194

single nucleotide polymorphism (SNP), 34

Skylab, 84

solar, 24, 29, 42, 54–55, 57–58, 84, 89–90, 95, 98, 100, 102, 104, 120–23, 142, 170–71, 174, 183, 204, 224–25, 232, 234, 238–39; cell, 101, 103, 121–23, 142; collector, 84, 120, 123; Heliospheric Observatory (SOHO), 24, 99; modulation, 174; particle event (SPE), 95, 174; power, 20, 42, 81, 89, 120, 122–24; radiation, 28, 29, 41, 171, 174–75, 183

Solar System, xiv, 6, 18–22, 25, 27, 37–38, 42–44, 47, 93, 98, 100, 103–4, 110, 119–20, 122, 143, 174, 183, 219–29, 234, 237–38

somatogravic illusion, 156

South Pole, 48, 69, 112, 124, 138, 207, 209, 228; Station, 90, 106, 124–25, 233

Soviet Union, 4, 45, 57, 70–71, 73, 83, 85, 182

Soyuz, xii, 15, 28, 73, 76, 83–87, 91

Space Age, 1, 45, 58, 73, 118

spaceflight, 24, 27, 40, 68

Space Launch System (SLS), 23, 191, 205

space medicine, 5, 62, 65, 66, 67

Space Race, 4, 14, 46, 70, 73

Space Shuttle, xi, 1, 23, 74, 196, 204

spacesuit, 70, 74, 95, 101, 104–6, 179, 183, 191, 212–14, 227, 234

station, xi, 1, 13, 74, 79, 80–85, 91, 138

SpaceX, 15

spallation, 184

spinoff, 4, 5, 45, 91, 113, 169, 170

Sputnik, 13, 57, 58, 70, 73

Stafford Task Group, 78

Stafford, T. P., 72

stars, 22, 25, 27, 37, 38, 44, 109, 143, 219–20, 224–25, 236–50

Stardust Probe, 123

Stefan-Boltzmann law, 232

Stern, S. A., 17, 96, 128

stochastic, 176, 180

stratosphere, 5

stress, 18, 30–32, 36, 39, 62, 77, 154, 166–68, 195

stressors, 31–32

Strughold, H., 65–66

submarines, 26, 29, 59–61, 106, 121, 123–24, 142–43, 146, 170, 225, 238

Sun, 9, 20, 24, 27, 37, 38, 41–43, 47, 99, 104, 109, 119–23, 142, 175, 183, 189–90, 207, 211, 215, 220–26, 229, 231–33, 237–45

superconductivity, 234

supernova, 174, 244, 246

surface technologies, 24, 89, 113, 119, 188–89, 191, 203, 211, 213, 221

taikonaut, 35, 91, 92

Tau Ceti, 239

Tegetthoff, 50

teleportation, 104

temperature, 6, 8, 9, 21, 28–30, 38, 42, 72, 75, 77, 84, 86, 89, 94, 96, 97, 102, 106–9, 112, 119–21, 133, 206–11, 214–15, 220, 224, 227–35, 238, 241, 243–45; body, 69, 108, 137, 247; surface, 38, 41, 207, 209, 230

Tereshkova, V., 74

thermodynamics, 28, 109, 120, 125, 232; efficiencies, 120, 142, 248

thyroid hormone, 167–68
Tibetans, 33, 34, 35, 172
Titan, 6, 19, 36, 42, 220, 225, 229–31, 243
Titania, 220, 231
titanium, 92, 99, 100, 102–3, 212
Tokomak, 22, 102
trait, 36, 39, 237
Triton, 42, 220, 231–32
troposphere, 29, 230
Tsiolkovsky, K., 81, 154
tuberculosis, 51, 52
tunneling, 104, 213
twilight, 18, 36, 149, 151, 200; winter, 167

ultraviolet, 5, 148; light, 95, 148, 178; radiation, 5, 175; spectrograph, 98
Uranus, 36, 122, 153, 220, 225, 231
urea, 245
USSR, 4, 69, 73

vacuum, xiv, 24, 44, 46, 68, 82, 94, 101–2, 105, 112, 195, 212, 232
Van Allen belts, 17–18, 22, 57–58, 80, 95, 170, 172–4, 183–84, 188, 206, 226
Van Allen, J., 57
Venus, 7, 38, 41, 122, 153, 189, 220, 240, 242
vestibular labyrinth, 155
vitamins, 135–36, 149–51; vitamin A, 149, 187; vitamin B, 149; vitamin C, 149; vitamin E, 149, 187; vitamin D, 136, 149, 150–51, 158, 163; vitamin K, 149, 151
Von Braun, W., 73, 83
Von Neumann, J., 44, 249, 250
Von Payer, J., 49, 50

water (H_2O), xiv, 5, 7, 9, 16, 19, 22, 27–31, 38–39, 41–43, 50, 52, 59–61, 66, 69, 80, 86, 90, 94–101, 104, 108–9, 113, 117, 121, 124–25, 127–49, 153, 162, 170, 184–85, 191, 194–97, 205–6, 208–12, 215, 222–24, 227–32, 238, 240, 242, 246
watt, 20, 100, 119, 122, 137, 232–33
weightlessness, 36, 39, 65, 70, 154–55, 157, 165, 198, 204
Weyprecht, K., 49–52
white blood cell, 165, 178
White, E. H., 70–71
Wilmer, W. H., 61
World War I, 61, 121
World War II, 45, 58–59, 62–63, 67, 73, 83
Wright Field, 62–65

xenon (Xe), 123, 224, 229, 231
X-rays, *173*, 176, 177, 180, 184

Young, J. W., 72

zeolite, 148
zero g, 82